Primal Proofs

Proofs Offering a Clearer Picture of Prime Numbers

By

Shelton W. Riggs, Jr.

Primal Proofs

**Proofs Offering a
Clearer Picture of Prime Numbers**

Copyright © 2007 by Shelton W. Riggs, Jr

All rights reserved. No part of this book may be used or reproduced by any means, graphic, electronic, or mechanical, including photocopying, recording, taping or by any information storage retrieval system without the written permission of the publisher except in the case of brief quotations embodied in critical articles and reviews.

ISBN-13: 978-1-44862-125-5

ISBN-10: 1-44862-125-9

This book is dedicated to all who are curious about prime numbers

About The Author

Shelton W. Riggs, Jr. earned undergraduate (University of Texas) and graduate (Vanderbilt) degrees in both Physics and Mathematics.

Professionally, he has consulted as both a hardware and software design engineer to numerous Fortune 500 companies for a wide range of scientific applications. He helped solve several scientific problems for US Army, Air Force and Navy.

Other interests include theoretical physics including quantum mechanics, relativistic mechanics and theoretical mathematics (especially the mystery of prime numbers).

Hobbies include dancing, karaoke, juggling, playing keyboards, writing songs, and writing poetry.

Other Works By Author

The Scientific Theory of God – A bridge Between Faith and Physics provides the reader with basic scientific understanding, interpretation, clarification and answers about concepts and beliefs associated with a Supreme Being. These ideas are developed and based on current theory and the standard model of physics. This new basis has revealed surprising relationships between the scientific definitions of both God and man. A model for the behavior of living matter (bioenergy) has been extended to include the behavior of human beings in terms of perception, decision and action. These concepts combined with the operation of short and long-term memory explain both human consciousness and how the mind controls the body. This model also includes how any desired behavior (provided it does not go against survival) may be achieved. This book offers a scientific creation theory and shows how it is compatible with both the big bang as well as evolutionary theory.

Nature of the First Cause – The Discovery of What Triggered the Big Bang contains the formal scientific theory of how the universe got started. It lays down the mathematical foundation for the creation theory put forth in "The Scientific Theory of God" book. It resolves the asymmetry problem of physics. It solves the two main cosmological problems by identifying both dark energy and dark matter. This theory predicts the correct order of magnitude for the number of galaxies and stars in the universe revealed by the Hubble ultra deep field results. It uncovers two entangled parallel worlds consisting of negative antimatter and positive matter. It explains the accelerated expansion of both matter and negative antimatter. It predicts the distance between matter and negative antimatter to be the Schwarzschild diameter of the expanding universe.

An Alternate Lorentz Invariant Relativistic Wave Equation offers an invariant form which differs from both the Dirac equation as well as the Klein-Gordon equation. Unlike, Schrodinger's non-relativistic wave equation, both the Dirac as well as the Klein-Gordon equation predict wave functions which do not collapse when applied to free systems at rest. On the other hand, Schrodinger's equation predicts wave functions that do collapse when applied to free systems at rest. The alternate relativistic wave equation offered by the author follows Schrodinger's philosophy that if a free system is at rest, then it is a particle with a collapsed wave function.

The Origin of the Planck Mass, Planck Length and Planck Time presents solutions of a system composed of two identical photons which are trapped in each other's gravitational field. The solution applies to any pair of identical particles having zero rest mass. Two solutions were derived. One solution was found by treating two photons as point particles. The quantum mechanical solution came about by treating the two photons as waves. In both solutions, the predicted distance between photons was found to be proportional to the Planck length. The period of the photon's orbit was proportional to the Planck time and the mass energy of each photon is proportional to the Planck mass.

The concept of a Planck length, Planck mass and Planck time all emerge from this single model.

The Nuclear Force shows how quark trios are bound into protons and neutrons.

Lines of Reflection include poems about life, love and death.

Acknowledgments

I acknowledge God for providing all the resources necessary to answer a few questions about prime numbers.

I acknowledge my country for providing me with all my freedoms especially, freedom of speech.

I acknowledge my parents for providing a secure and nurturing environment that initially made my learning fun.

I acknowledge all my teachers especially my science and mathematics teachers.

I acknowledge every author in the reference section of this book for providing both ideas and data.

Preface

This study was undertaken to investigate the truth of several conjectures and hypothesis related to prime numbers. This includes Christian Goldbach's (binary and ternary) conjectures as well as the Riemann hypothesis. Proofs of whether or not there is an unlimited number of prime triplets (three primes whose consecutive difference is two (2)) and the twin prime conjecture (that there is an unlimited number of twin primes (two primes whose difference is two (2)) are presented. Proof of both Anglica's conjecture (difference between the square roots of two consecutive primes is less than unity) and Legendre's conjecture (there is a prime between the squares of any two consecutive natural numbers) are presented. Proof of Brocard's conjecture that there are at least four (4) primes between the squares of any two consecutive odd primes is offered.

The initial form of the prime number theorem (PNT) as Gauss's first prime number conjecture is used to create a table that demonstrates the truth of the binary conjecture and serves as a model for its proof. A formula is deduced that yields a minimum number of ways in which any even number can be expressed as the sum of two primes.

A proof of Goldbach's binary conjecture (that every even natural number greater than two (2) can be expressed as the sum of two (2) primes) is given. The proof is by contradiction.

A proof of Goldbach's ternary conjecture (that all natural numbers greater than or equal to six (6) is the sum of three (3) primes) is presented. The proof of Goldbach's binary conjecture is used in the proof of Goldbach's ternary conjecture. A proof that any even number greater than twelve (>12) satisfies the binary conjecture in a plurality of ways.

A proof that there is only one set of prime triplets (any three consecutive primes that differ by two (2)) and a proof that there is only one set of prime singlets (two primes that differ by one (1)) is presented. These proofs were included for completeness.

A demonstration of the twin prime conjecture (that there is an infinite number of twin primes (any two primes that differ by two (2)) is developed. The initial form of the prime number theorem (PNT) as Gauss' first prime number conjecture is utilized in the demonstration. A proof by construction (utilizing the proof of the binary conjecture) of the twin prime conjecture is offered. A proof by deduction that any prime greater than three (3) is the average of two (2) other odd primes is included.

Proofs of the Riemann hypothesis by deduction and contradiction are offered. An alternate (besides Euclid's proof by contradiction) proof that there is an unlimited number of primes is developed using the PNT.

How all primes greater than three (3) fall into two entangled categories is presented.

Comprehensive formulas for all the sequences of odd composites are developed.

Two entangled formulas that generate all the primes beyond the second prime ($P_1 = 2$, $P_2 = 3$) are developed and summarized. An example is given of how the third prime which is $P_3 = 5$ can generate the next fourteen (14) primes.

A glossary is provided which lists and explains the fundamental theorems, relationships, conjectures, functions and rules associated with or related to prime numbers. Several of these theorems, rules and functions have been utilized for the proofs in this manuscript.

Additionally, the Peano axioms and Euclid's axioms are listed and described. The Peono axioms form the foundation of all natural numbers. Euclid's axioms are the foundation for Euclidean geometry and the meaning of equality.

The fundamental rules of mathematics (which can be deduced from the Peano and Euclid's axioms) are also provided. These rules offer relational definitions which form the basis of algebra, trigonometry, hyperbolic functions, complex numbers and the calculus.

Additionally, the five types of known numbers are defined and explained.

Table of Contents

Leading Page	i-ii
Title Page	ii
Copyright Page	iii
Dedication	iv
About the Author	v
Other Works By Author	vi
Acknowledgments	x
Preface	xi
Table of Contents	xv
Chapter 1 Conventions	1
Chapter 2 Tools	4

Chapter 3 Definitions 8

**Chapter 4 Analysis of Binary Conjecture
      via Prime Number theorem** 11

Chapter 5 Proof of Binary Conjecture 20

**Chapter 6 Proofs of Ternary Conjecture
and that Evens Satisfy Binary Conjecture in
Plurality of Ways** 33

**Chapter 7 Proof of Limited Prime Triplets and
Singlets plus Proofs of Andrica's, Legendre's
and Brocard's Conjectures** 42

**Chapter 8 Proofs of Twin Prime Conjecture
and that Prime > 3 is Average of Two Other
Odd Primes** 55

**Chapter 9 Proof of Riemann Hypothesis
and Alternate Proof of Unlimited Number
of Primes** 68

Chapter 10 Prime Entanglement   83

Chapter 11 Odd Composites   91

Chapter 12 Prime Number Generator   108

Table 1   17

Table 2   43, 84

Table 3   106

Glossary   121

Axiomatic Foundations   139

Fundamental Rules of Mathematics   144

Figure 1   149

Types of Numbers   155

References   157

Index 165

Trailing Pages 173-176

Chapter 1

Conventions

In this study, the following conventions have been adopted:

(C1) A, B, C, E, I, J, K, L, M, N, O, Θ, P, Q, S, T, U, X, Y, Z all stand for natural numbers. P stands for a prime number. O stands for an odd natural and E stands for an even natural. Θ stands for an odd composite. U_0 stands for unity (1).

(C2) i, j, k, l, m, n, q, r, s, all stand for natural numbers usually used for indices. Most indices will be formulated to start with one (1).

2 Primal Proofs

(C3) E_J stands for the Jth even natural number.
$(E_J = 2J, J = 1, 2, 3 \ldots)$

(C4) P_J stands for the Jth prime number.
$(P_1 = 2, P_2 = 3, P_3 = 5, \text{etc..})$

(C5) O_J stands for the Jth odd natural number.
$(O_J = 2J - 1, J = 1, 2, 3 \ldots)$

(C6) N_J stands for the Jth natural number.
$(N_1 = 1, N_2 = 2, N_3 = 3, \text{etc..})$

(C7) D, F, R, X, Y, Z, x, y, z, & Ψ all stand for real numbers. Note X, Y, Z may also be natural indices.

(C8) $Ln(X)$ means the natural logarithm of X.

(C9) $\approx$ means is approximately equal to.

(C10) $\{B_1, B_2, \ldots B_J \mid j \neq 0\}$ means a sequence of natural numbers (B_J is the Jth natural) further specified by what follows the | symbol.

(C11) ∈ means is a member of what follows. An example is $B_1 \in \{B_1, B_2, \ldots B_J \mid j \neq 0\}$

(C12) ∉ means is not a member of what follows. An example is $B_0 \notin \{B_1, B_2, \ldots B_J \mid j \neq 0\}$

(C13) $\prec R \succ$ means the integer part of any real number R.

(C14) In this manuscript, both brackets [ ] and parenthesis ( ) will (depending on context) denote factors. The parenthesis in f(x) means function of x.

(C15) All the natural numbers including unity, odds, evens and primes are enumerated as follows:

```
1    2    3    4    5    6    7    8    9    10   11   12   13 ...
N₁   N₂   N₃   N₄   N₅   N₆   N₇   N₈   N₉   N₁₀  N₁₁  N₁₂  N₁₃
O₁        O₂        O₃        O₄        O₅        O₆        O₇ ...
     E₁        E₂        E₃        E₄        E₅        E₆        ...
U₀  P₁   P₂        P₃        P₄                       P₅        P₆ ...
```

Chapter 2

Tools

The following tools have been utilized. The prime counting function, $\pi(N)$ is the actual number of primes less than or equal to some natural number, N. The symbol $\pi(N)$ is not to be confused with the circumference of a circle divided by its diameter. Mathematicians have assigned this symbol to be the prime counting function.

The form of the prime number theorem (PNT) utilized in this study is

(T1) $\pi(N) \approx N/Ln(N)$

The PNT utilizing the form of equation (T1) guarantees that

(T2) $N/Ln(N) < \pi(N)$

and since by the Rosser - Schoenfeld (RS) Theorem that

(T3) $N/Ln(N) < \pi(N) < 1.25506 N/Ln(N)$

where N > 16 (See reference Rosser & Schoenfeld 1962), the prime number theorem as expressed by equation (T1) is the form that will be utilized.

An equivalent statement of this form of the prime number theorem is that the probability, $\Psi(N)$ that N is prime is

(T4) $\Psi(N) \approx 1/Ln(N)$

Note that an estimate of the number of primes in N naturals is to multiply the probability of N being

6 Primal Proofs

a prime (1/Ln(N)) times the number of naturals (N) or N/Ln(N) which is precisely the PNT.

Riemann's Extended Zeta Function, $\zeta(s)$ is

(T5) $\zeta(s) = 2^s \pi^{s-1} \mathrm{Sin}(\pi s/2) \Gamma(1-s) \zeta(1-s)$

where $s \neq 1$ is complex, Sin is the normal trigonometric sine function and Γ is the gamma function. See Glossary and Rules of math section.

L'Hopital's rule: If two functions $f(x)$ and $g(x)$ at the point $x = a$ are either both zero ($f(a) = g(a) = 0$) or infinite, ($f(a) = g(a) = \infty$), then

(T6) $\lim_{x \to a} f(x)/g(x) = \lim_{x \to a} (df/dx)/(dg/dx)$

(T7) Mod(A, B) = remainder of A/B where A and B are both natural numbers and $B \neq 0$.

(T8) $A^Z = A^{X+iY} = A^X[\cos(\mathrm{Ln}A^Y) + i\mathrm{Sin}(\mathrm{Ln}A^Y)]$

where A is a real, Z is complex and $i = (-1)^{1/2}$

The Laurent series expansion of the Reimann zeta (see glossary) function is:

(T9) $\zeta(s) = 1/(s-1) + \gamma_0 + \gamma_1(s-1) + \gamma_2(s-1)^2 + \ldots +$

where $s \neq 1$. The constants, γ_k here are called the Stieltjes constants and defined as

(T10) $\gamma_k = (-1)^k/k! \lim_{N \to \infty} (\sum_{m \leq N} Ln^k(m)/m - Ln^{k+1}(N)/(k+1))$

The first constant term γ_0 is also called the Euler-Mascheroni constant. See reference
http://www.exampleproblems.com/wiki/index.php/Riemann_zeta_function

Derichlet's eta functiion, $\eta(s)$, (see glossary) is related to the Reimann zeta function $\zeta(s)$ as:

(T11) $\zeta(s) = \eta(s)/(1 - 2^{-(s+1)})$

extending the domain of the zeta function to $s > 0$, and $s \neq 1$ where s is in general a complex number.

Chapter 3

Definitions

Preliminary definitions are as follows. By the prime number theorem (PNT), the number of primes $\leq E_J$ (where E_J is the jth even number) is given by

(D1) $\pi(E_J) \approx E_J/Ln(E_J)$

where $\pi(E_J)$ is the conventional designation for the actual number of primes less than or equal to E_J.

Define an approximate number of primes less than or equal to E_J as

(D2) $N_{JP} \approx \langle E_J/Ln(E_J) \rangle$

Definitions 9

Recall that in this study, the enclosure $\triangleleft R \triangleright$ means "integer part of R".

Define the sequence

(D3) $\{P_K, \mid K = 1, 2, 3, \ldots \pi(E_J)\}$

as the set of $\pi(E_J)$ primes $< E_J$. When $K = 1$, define

(D4) $O_{J1} = E_J - P_1 + 1 = E_J - 2 + 1 = E_J - 1$

where $P_1 = 2$ and O_{J1} yields the odd number which is one less than E_J. For odd primes, P's > 2, define

(D5) $\{O_{JK}\} = \{E_J - P_{K+1} \mid K = 1, 2, 3, .. \pi(E_J) - 1\}$

Note that sequence (D5) generates $\pi(E_J) - 1$ unique monotonically decreasing set of odd natural numbers, one for each successive prime in the range $2 < P_{K+1} < E_J$. Recall by the prime number theorem (PNT), that definition

(D2) $N_{JP} \approx \lessdot E_J/Ln(E_J) \gtrdot$

is the estimated total number of primes less than or equal to E_J including $P_1 = 2$. Recall that the exact number of odd primes less than E_J is $\pi(E_J) - 1$ (since $\pi(E_J)$ includes $P_1 = 2$). The (D5) sequence produces $\pi(E_J) - 1$ unique odds (all $< E_J$) and all coming from the set of odds $< E_J$.

Define an operator Ezsort as

(D6) Ezsort $\{C, A, B, 0, B, C, 0, B, A\} = \{A, B, C\}$

where $A < B < C$. Ezsort operates on a sequence of non negative integers and produces a new sorted ascending sequence of non repeated positive natural numbers containing no zeros.

(D7) If $s = x + iy$, then $Re(s) = x$ and $Im(s) = y$

where $i = (-1)^{1/2}$

Chapter 4

Proof of Binary Conjecture via Prime Number Theorem

What follows offers an analysis via the prime number theorem (PNT) (Gauss' first estimate not involving the logarithmic integral) of a unique set of odd numbers. These odds are generated by subtracting each odd prime (beneath an even number) from that even number. The resulting table (Table1) will be used to demonstrate the plurality of ways the Goldbach binary conjecture is satisfied for a given even number.

12 Primal Proofs

Appropriate formulas are developed for each column of the table.

The number of primes in a run of naturals $< E_J$ is estimated by equation

(D2) $N_{JP} \approx \lessgtr E_J/Ln(E_J) \gtrless$

and thus, the frequency of primes is

(AB1) $\Psi_P \approx N_{JP}/E_J \approx 1/Ln(E_J)$

Note this is also compatible with (T4) $\Psi(N) \approx 1/\ln(N)$. Definition

(D5) $\{O_{JK}\} = \{E_J - P_{K+1} \mid K = 1, 2, 3, .. \pi(E_J) - 1\}$

produces $\pi(E_J) - 1$ unique odds all $< E_J$ and all coming from the set of odds $< E_J$. It will be shown that the set of odds generated by equation (D5) are distributed the same as for a run of odd naturals generated by

(D6.1) $\{O_L\} = \{2L - 1 \mid L = 1, 2, 3, ..\}$

Minimum Number of Ways to Satisfy Binary Conjecture

Thus, a minimal number of odd primes $< E_J$ is estimated to be $(N_{JP} - 1)$ (since equation (D2) is assumed to have included $P_1 = 2$). This is also the number of odds generated by equation (D5). Therefore, the number of primes in $\{O_{JK}\}$ is estimated as $(N_{JP} - 1)$ multiplied by the frequency of primes or

(AB2) $N_{OP} \approx \prec(N_{JP} - 1)\Psi_P\succ$

$\approx \prec(E_J - Ln(E_J))/(Ln^2(E_J))\succ$

Moreover, if $N_{OP} = 2$, then the number of ways in which two primes add to E_J is only one (1), because if

(AB3) $E_J - P_A = P_B$

then also

(AB4) $E_J - P_B = P_A$

which is the same as equation (AB3). Thus, a minimum (by ignoring cases where $E_J = 2P_C$) number of ways in which two primes add to E_J is

(AB5) $N_W = \langle (N_{OP})/2 \rangle$

Table 1 Description

The 1st column of table 1 is $\pi(E_J)$, the actual number of primes $\leq E_J$.

The 2nd column is equation

(D2) $N_{JP} \approx \langle E_J/Ln(E_J) \rangle$

which gives an approximate number of primes less than E_J and less than the actual number $\pi(E_J)$.

The 3rd column is the actual number of odd primes in each E_J row including all the elements to the right of column O_{J1}.

The 4rth column is equation

(AB2) $N_{OP} \approx \prec (E_J - Ln(E_J))/(Ln^2(E_J)) \succ$

which is a minimal number of odd primes in each Ej row (to the right of O_{J1}).

The 5th column is equation

(AB5) $N_W = \prec N_{OP}/2 \succ$

which is the minimum number (ignoring cases where $E_j = 2P_C$) of ways in which 2 odd primes can add up to the given E_J. Note that the first non-zero N_W (not counting 1st row result) is for $E_J = 24$.

The 6th, 7th, 8th and 9th columns are the naturals, odd primes, odds and evens respectively.

The 10th column is $O_{J1} = E_J - 1$. The elements of each E_J row (to the right of the O_{J1} column) was generated by subtracting $(\pi(E_J) - 1)$ odd primes

16 Primal Proofs

from the E_J belonging to that row. The sequence of odd naturals in E_J's row was given by equation

(D5) $\{O_{JK}\} = \{E_J - P_{K+1} \mid K = 1, 2, 3, .. \pi(E_J) - 1\}$

Note that every O_{JK} column begins with unity since for that particular E_J there exists the largest odd prime, $P_{K+1} < E_J$ such that

(AB5.1) $E_J - P_{K+1} = 1$

Thus, each O_{JK} column to the right of the O_{J1} column (O_{J2}, O_{J3}, O_{J4}, O_{J5}, ...) evolves into a sequential run of all odd naturals as if generated by

(D6.1) $\{O_L\} = \{2L - 1 \mid L = 1, 2, 3, ..\}$

Therefore, every element of $\{O_{JK}\}$ belongs to a unique sequential run of odd naturals, $O_{JK} \in \{O_L\}$. Hence, $\{O_{JK}\}$ must have the same distribution of odd primes as in a run of odd naturals or for that matter, in a sequential run of naturals greater than two (2).

Proof of Binary Conjecture via PNT

$\pi(E_j)$	N_{JP}	π(E's Row)	N_{OP}	N_W	N_J	P_{J+1}	O_J	E_J	O_{J1}	O_{J2}	O_{J3}	O_{J4}	O_{J5}	O_{J6}	O_{J7}	O_{J8}	O_{J9}...
1	2	0	2	1	1	3	1	2	1								
2	2	0	1	0	2	5	3	4	3	1							
3	3	1	1	0	3	7	5	6	5	3	1						
4	3	2	1	0	4	11	7	8	7	5	3	1					
4	4	3	1	0	5	13	9	10	9	7	5	3					
5	4	2	1	0	6	17	11	12	11	9	7	5	1				
6	5	3	1	0	7	19	13	14	13	11	9	7	3	1			
6	5	4	1	0	8	23	15	16	15	13	11	9	5	3			
7	6	4	1	0	9	29	17	18	17	15	13	11	7	5	1		
8	6	4	1	0	10	31	19	20	19	17	15	13	9	7	3	1	
8	7	5	1	0	11	37	21	22	21	19	17	15	11	9	5	3	
9	7	6	2	1	12	41	23	24	23	21	19	17	13	11	7	5	1
9	7	5	2	1	13	43	25	26	25	23	21	19	15	13	9	7	3
9	8	4	2	1	14	47	27	28	27	25	23	21	17	15	11	9	5
10	8	6	2	1	15	53	29	30	29	27	25	23	19	17	13	11	7 1
11	9	4	2	1	16	59	31	32	31	29	27	25	21	19	15	13	9 3
11	9	7	2	1	17	61	33	34	33	31	29	27	23	21	17	15	11 5
11	10	8	2	1	18	67	35	36	35	33	31	29	25	23	19	17	13 7
12	10	3	2	1	19	71	37	38	37	35	33	31	27	25	21	19	15 9
12	10	6	2	1	20	73	39	40	39	37	35	33	29	27	23	21	17 11
13	11	8	2	1	21	79	41	42	41	39	37	35	31	29	25	23	19 13
14	11	6	2	1	22	83	43	44	43	41	39	37	33	31	27	25	21 15
14	12	7	2	1	23	89	45	46	45	43	41	39	35	33	29	27	23 17

Table 1

18 Primal Proofs

The 11th column beginning with O_{J2} and all columns thereafter (O_{J2}, O_{J3}, O_{J4}, ...) evolve into a set of all the odd naturals as each E_J row is filled in by elements generated by the prescription

(D5) $\{O_{JK}\} = \{E_J - P_{K+1} \mid K = 1, 2, 3, .. \pi(E_J) -1\}$

A minimal number of ways that two odd primes add up to an even number (a derivation based on the prime number theorem) is demonstrated (by table 1) for any even natural number larger than or equal to 24 (since the minimal number of ways, $N_W \geq 1$ for $E_J \geq 24$ ignoring the case when $E_J = 2$). This means there must be at least one (1) prime in the corresponding set of odds in that E_J row. Therefore, for some $K = Q$ in that E_J row, we have element $O_{JQ} = P_{Q+1}$. However, the element of equation (D5) is

(AB5.2) $O_{JK} = E_J - P_{K+1}$

and O_{JQ} must have resulted from subtracting some prime $P_{M+1} < E_J$, from E_J. Thus,

(AB6) $O_{JQ} = P_{Q+1} = E_J - P_{M+1}$

and when rewritten becomes

(AB7) $E_J = P_{Q+1} + P_{M+1}$

which demonstrates the binary conjecture for $E_J \geq 24$. For $6 \leq E_J \leq 24$, it is well known that $E_J = P_{X+1} + P_{Y+1}$ where $P_{X+1} + P_{Y+1}$ are both odd primes. Note that $E_J = 4 = P_1 + P_1 = 2 + 2$. Thus, the binary conjecture is proven (by the above equations and definitions) that generate table 1 columns for all $E_J > 2$. Q.E.D.

Chapter 5

Proof of Goldbach's Binary Conjecture

Goldbach's binary conjecture will now be proven. Reference will be made to table 1, especially the elements to the right of the evens column.

Even numbers, $E_J \leq 10^{14}$ have now been tested to see if they are comprised of two primes (please refer to Goldbach's binary conjecture in the glossary for references) with no binary conjecture contradiction.

Assumptions

Assume that there exists some big even natural, E_B (as may be invisioned by $E_B = 46$ at the bottom of table 1) that cannot be expressed as the sum of two primes. In other words, every odd prime less than this big even number ($P_{J+1} < E_B$) that is subtracted from E_B results in an odd composite. Note that this means

(PB1) $E_B - P_{\pi(E)} \neq 1$

where $P_{\pi(E)}$ is the largest prime less than this big even number E_B.

Inequality (PB1) must hold, since it could be argued that unity, U_0 (1) meets the requirements of primality (i.e. defining a zeroth prime as $P_0 = U_0 = 1$). Thus, if equation (PB1) is not true, then it would mean that $E_B = P_{\pi(E)} + P_0$ which would technically violate the initial assumption that E_B cannot be so expressed. Thus, column O_{J1} should be viewed as the initial column of all odd naturals, $E_J - 1$.

Subtracting unity (1) from E_B generates the largest, odd composite element of column $O_{J1} < E_B$

(PB1.1) $E_B - 1 = \Theta_0$

where Θ_0 is the largest odd composite element in the run of odd naturals of column $O_{J1} < E_B$.

Start of Proof

Subtracting all of E_B's odd primes ($P_{j+1} < E_B$) from E_B yields the following system of $[\pi(E_B)-1]$ equations (PB2) to (PB5). This also presumably generates E_B's giant row of pure odd composites. Moreover, it also must add one more element to all the $[\pi(E_B)-1]$ existing columns of odd naturals. Furthermore it cannot start a new column since this would imply

(PB1.2) $E_B - P_{\pi(E)} = 1$

which has been ruled out by equation (PB1) above. Thus, the above subtraction process yields

(PB2) $E_B - P_2 = \Theta_1$

(PB3) $E_B - P_3 = \Theta_2$

....

....

(PB4) $E_B - P_{J+1} = \Theta_J$

....

....

(PB5) $E_B - P_{\pi(E)} = \Theta_{\pi(E)-1}$

Θ_J means the Jth odd composite and again, $P_{\pi(E)}$ is the largest prime $< E_B$. Note that the subscript $\pi(E)$ has been shortened to mean

(PB6) $\pi(E) = \pi(E_B)$

It will be instructive to note that all the odd primes that were subtracted from E_B came from column O_{J1} which contains $\pi(E_B)-1$ odd primes. Note also, that the smallest odd composite is nine (9). Summing all the equations (PB2) through (PB5) yields

24 Primal Proofs

$$(PB7) \sum_{J=1}^{\pi(E_B)-1}(E_B - P_{J+1}) = \sum_{J=1}^{\pi(E_B)-1}\Theta_J$$

which reduces to

$$(PB7.1) \sum_{J=1}^{\pi(E_B)-1}\Theta_J = [\pi(E_B)-1]E_B - \sum_{J=1}^{\pi(E_B)-1}P_{J+1}$$

The left side of this equation represents the sum of all the $[\pi(E_B) -1]$ assumed odd composites in E_B's row. Referring to table 1, the last odd composite column element of O_{J2} (the second column after the Evens column) is

(PB8) $2N_O - 1 = E_B - P_2 = E_B - 3$

where N_O is the total number of odd elements in the O_{J2} column since the O_{J2} column is also a run of odd naturals. Solving equation (PB8) for N_O yields

(PB9) $N_O = E_B/2 - 1$

Proof of Goldbach's Binary Conjecture 25

The sum of all the elements of the O_{J2} column is N_O^2 because of the well known fact that the sum of all the odd naturals from 1 to N is N^2. Thus,

$$(PB10) \quad \sum_{m=1}^{E_B/2-1}(2m-1) = (E_B/2-1)^2$$

The total number of odds, N_O, odd primes and odd composites in the O_{J2} column are related by

$$(PB10.1) \; N_O = N_C + \pi(E_B - 2) - 1 + 1$$
$$= N_C + \pi(E_B - 2)$$

where N_C is the number of odd composites and $[\pi(E_B - 2) - 1]$ is the number of odd primes in column O_{J2} plus one for the odd unity (1) element which starts column O_{J2}. Utilizing equation (PB9), there is a grand total of

$$(PB10.2) \; N_C = E_B/2 - 1 - \pi(E_B - 2) =$$
$$E_B/2 - [\pi(E_B - 2) + 1]$$

odd composites in the O_{J2} column. Note that the prime counting function $\pi(E_B - 2) - 1$ notation refers to the odd primes in column O_{J2}. However, recall that the last element of column O_{J1} was the odd composite

(PB1.1) $E_B - 1 = \Theta_0$

Column O_{J2} is the same as column O_{J1} with its last element (an odd composite) removed. Thus, there are exactly the same number of odd primes under E_B as there are under $E_B - 2$. Thus,

(PB10.3) $\pi(E_B - 2) = \pi(E_B)$

Rewriting equations (PB10.1) and (PB10.2) yields

(PB10.4) $N_O = N_C + \pi(E_B)$

and

(PB10.5) $N_C = E_B/2 - [\pi(E_B) + 1]$

The sum of all the odd naturals in the O_{J2} column is also unity (1) plus the sum of all $[\pi(E_B - 2) - 1] = [\pi(E_B) - 1]$ odd primes plus the sum of all the odd composites. Thus,

$$\text{(PB11)} \quad 1 + \sum_{J=1}^{\pi(E_B)-1} P_{J+1} + \sum_{k=1}^{E_B/2 - [\pi(E_B)+1]} \Theta_k = (E_B/2 - 1)^2$$

Solving (PB11) for the sum of the odd primes and substituting it into equation (PB7.1) yields

$$\text{(PB12)} \quad \sum_{J=1}^{\pi - 1} \Theta_J = (\pi - 1)E_B + 1 + \sum_{k=1}^{E_B/2 - (\pi+1)} \Theta_k - (E_B/2 - 1)^2$$

where the notation π has been further simplified to mean

(PB12.1) $\pi = \pi(E_B)$

Equation (PB12) can further be reduced to

$$\text{(PB13)} \quad \sum_{J=1}^{\pi - 1} \Theta_J = \pi E_B + \sum_{k=1}^{E_B/2 - (\pi+1)} \Theta_k - E_B^2/4$$

28 Primal Proofs

On the left of this equation is the sum of all E_B's row of odd composites. The second term on the right is the sum total of all the odd composites in the O_{J2} column.

E_B's row of odd composites, Θ_J on the left of equation (PB13) are all monotonically decreasing until the last and largest prime (P_π) has been subtracted from E_B and yielding the least odd composite, $\Theta_{\pi(E)-1}$ as specified by equation (PB5) and in view of (PB12.1) yields

(PB5.1) $E_B - P_\pi = \Theta_{\pi-1}$

The total number of odd composites on the right of equation (PB13), are monotonically increasing from nine (9) (the smallest odd composite) to the largest odd composite which must be (by assumption)

(PB13.1) $E_B - P_2 = E_B - 3$

(the last element of the O_{J2} column). Note that each Θ_J on the left of equation (PB13) belongs to the sequence of total column O_{J2} odd composites $< E_B$ on the right of equation (PB13) denoted by $\{\Theta_k\}$, or

(PB13.2) $\Theta_J \in \{\Theta_k \mid k = 1, 2, 3 \ldots E_B/2 - (\pi + 1)\}$

where the symbol, $\in$ means "is a member of". Thus, all the odd composites on the left (E_B's row of odd composites) of equation (PB13) will get canceled out (this is what causes the ensuing contradiction) by the total odd composites on the right of equation (PB13). Thus, equation (PB13) is reduced to

(PB14) $\pi E_B + \sum_{k=1}^{E_B/2 - 2\pi} \Theta_k - E_B^2/4 = 0$

and can be rewritten as

(PB15) $\pi E_B = E_B^2/4 - \sum_{k=1}^{E_B/2 - 2\pi} \Theta_k$

or

(PB16) $\pi = E_B/4 - \sum_{k=1}^{E_B/2 - 2\pi} \Theta_k/E_B$

which clearly violates the prime number theorem (PNT) which says that

(PB17) $\pi \approx E_B/Ln(E_B)$

Note that in equation (PB16) the kth fraction, $\Theta_k/E_B < 1$ for any k since each of the odd composites is less than E_B. For large E_B, each fraction is small. Nevertheless, the value on the right of equation (PB16) cannot exceed $E_B/4$ which says that the number of primes cannot exceed $E_B/4$. Thus, setting

(PB18) $\pi \approx E_B/4$

will yield the upper bound for the number of primes less than E_B according to (PB16). By the PNT, equation (PB18) is

(PB19) $E_B/Ln(E_B) \approx E_B/4$

reducing the above equation to

(PB19.1) $Ln(E_B) \approx 4$

or that

(PB20) $E_B \approx e^4 \approx 54$

which is absurd since it is known that even naturals in the vicinity of fifty-four (≈ 54) do not violate Goldbach's binary conjecture. Furthermore, even numbers in this vicinity (≈ 54) are very far from the assumption that the big even natural, E_B was supposed to be greater than 10^{14}.

This means the original assumption that all the elements in E_B's row are composites cannot be true. Therefore, if all the elements of E_B's row are not composite, and since the last element of E_B's row cannot be unity (1), at least one or more elements must be an odd prime. This means that for some P_{J+1} in the system of equations (PB2) to (PB5) one or more of the composites, Θ_J must be an odd prime, P_{m+1}, where m is a natural number, thus

(PB21) $\Theta_J = P_{m+1}$

and equation

(PB4) $E_B - P_{J+1} = \Theta_J$

is reduced to

(PB22) $E_B - P_{J+1} = P_{m+1}$

which when rewritten is

(PB23) $E_B = P_{m+1} + P_{J+1}$

which contradicts the original assumption and says that all evens > 4 can always be expressed as the sum of two (2) odd primes. Moreover, since

(PB24) $4 = 2 + 2 = P_1 + P_1$

it follows that all evens > 2 may be expressed as the sum of two primes. Q.E.D.

Chapter 6

Proofs of Goldbach's Ternary Conjecture and that Evens > 12 Satisfy Binary Conjecture in Plurality of Ways

The proof of Goldbach's ternary conjecture will now be given.

It has been shown that for any even natural number, $E_J > 4$, there is at least one way to express it as the sum of two odd primes. Consider any $E_J \geq 8$, then the previous even $(E_J - 2)$ can be expressed as

34 Primal Proofs

(PT1) $E_J - 2 = P_A + P_B = N_{2(J-1)}$

where $J \geq 4$. This means that E_J can be expressed as

(PT1.1) $E_J = P_A + P_B + 2 = P_A + P_B + P_1 = N_{2J}$

However, the next odd natural number after E_J is

(PT1.2) $E_J + 1 = P_A + P_B + 3 = P_A + P_B + P_2$
$= N_{2J+1}$

Note that E_J is also the sum of two primes say,

(PT2) $E_J = P_C + P_D$

so that

(PT3) $E_J + 2 = P_C + P_D + 2 = P_C + P_D + P_1$
$= N_{2J+2}$

Moreover, the next natural number after $E_J + 2$, is

(PT4) $E_J + 3 = P_C + P_D + 3 = P_C + P_D + P_2$
$= N_{2J+3}$

Thus, in summary, we have four consecutive naturals as the sum of three (3) primes, namely

(PT5.1) $N_{2J} = P_A + P_B + P_1$

(PT5.2) $N_{2J+1} = P_A + P_B + P_2$

(PT5.3) $N_{2J+2} = P_C + P_D + P_1$

(PT5.4) $N_{2J+3} = P_C + P_D + P_2$

Thus, it can be inductively shown that for any natural, $N_J \geq 8$ can be expressed as the sum of three (3) primes. For the rest of the cases, observe that:

(PT6) $7 = 2 + 2 + 3 = P_1 + P_1 + P_2$

(PT7) $6 = 2 + 2 + 2 = P_1 + P_1 + P_1$

which means that any natural $N_{J+5} \geq 6$ can be expressed as the sum of three (3) primes or

(PT8) $N_{J+5} = P_X + P_Y + P_Z$

where J = 1, 2, 3, …

and for J > 1, either

(PT9) $P_Z = P_1 = 2$

for even naturals or

(PT10) $P_Z = P_2 = 3$

for odd naturals proving Goldbach's ternary conjecture. Q.E.D.

Proof Evens Greater than Twelve (E >12) Satisfy Binary Conjecture in a Plurality of Ways

In chapter 4, it was demonstrated by the PNT, that any even number greater than or equal to twenty-four (24) satisfies Goldbach's binary conjecture. Equations

(AB2) $N_{OP} \approx \lfloor (E_J - Ln(E_J))/(Ln^2(E_J)) \rfloor$

and

(AB5) $N_W = \lfloor N_{OP}/2 \rfloor$

express that N_{OP} is the number of odd primes resulting from subtracting all odd primes less than some even number, E_J from that even number. N_W is a minimal number of ways that two primes add up to E_J. Both these formulas were constructed via the PNT. It turned out that N_W was greater than zero for E_J greater than or equal to twenty four (24).

It also turns out that by equations (AB2) and (AB5) that for any even number greater than or equal to eighty four (84), there are at least two ways in which the binary conjecture holds. For the rest of the cases (evens less than eighty four (84)) observe that

38 Primal Proofs

(PW1) $82 = 23+59 = 29+53 = P_9 + P_{17} = P_{10} + P_{16}$

(PW2) $80 = 37+43 = 19+61 = P_{12} + P_{14} = P_8 + P_{18}$

(PW3) $78 = 31+47 = 19+59 = P_{11} + P_{15} = P_8 + P_{17}$

(PW4) $76 = 29+47 = 23+53 = P_{10} + P_{15} = P_9 + P_{16}$

(PW5) $74 = 7+67 = 13+61 = P_4 + P_{19} = P_6 + P_{18}$

(PW6) $72 = 11+61 = 13+59 = P_5 + P_{18} = P_6 + P_{17}$

(PW7) $70 = 23+47 = 17+53 = P_9 + P_{15} = P_7 + P_{16}$

(PW8) $68 = 31+37 = 7+61 = P_{11} + P_{12} = P_4 + P_{18}$

(PW9) $66 = 5+61 = 7+59 = P_3 + P_{18} = P_4 + P_{17}$

(PW10) $64 = 5+59 = 11+53 = P_3 + P_{17} = P_5 + P_{16}$

(PW11) $62 = 3+59 = 19+43 = P_2 + P_{17} = P_8 + P_{14}$

(PW12) $60 = 7+53 = 13+47 = P_4 + P_{16} = P_6 + P_{15}$

(PW13) $58 = 11+47 = 17 +41 = P_5 + P_{15} = P_7 + P_{13}$

(PW14) $56 = 19+37 = 13 +43 = P_8 + P_{12} = P_6 + P_{14}$

(PW15) $54 = 11+43 = 7 +47 = P_5 + P_{14} = P_4 + P_{15}$

(PW16) $52 = 11+41 = 23 +29 = P_5 + P_{13} = P_9 + P_{10}$

(PW17) $50 = 7+43 = 13 +37 = P_4 + P_{14} = P_6 + P_{12}$

(PW18) $48 = 7+41 = 11 +37 = P_4 + P_{13} = P_5 + P_{12}$

(PW19) $46 = 3+43 = 17 +29 = P_2 + P_{14} = P_7 + P_{10}$

(PW20) $44 = 7+37 = 13 +31 = P_4 + P_{12} = P_6 + P_{11}$

(PW21) $42 = 11+31 = 5 +37 = P_5 + P_{11} = P_3 + P_{12}$

(PW22) $40 = 11+29 = 17 +23 = P_5 + P_{10} = P_7 + P_9$

(PW23) $38 = 19+19 = 7 +31 = P_8 + P_8 = P_4 + P_{11}$

(PW24) $36 = 13+23 = 7 +29 = P_6 + P_9 = P_4 + P_{10}$

40 Primal Proofs

(PW25) $34 = 3+31 = 11 + 23 = P_2 + P_{11} = P_5 + P_9$

(PW26) $32 = 3+29 = 13 + 19 = P_2 + P_{10} = P_6 + P_8$

(PW27) $30 = 13+17 = 11 + 19 = P_6 + P_7 = P_5 + P_8$

(PW28) $28 = 11+17 = 5 + 23 = P_5 + P_7 = P_3 + P_9$

(PW29) $26 = 3+23 = 7 + 19 = P_2 + P_9 = P_4 + P_8$

(PW30) $24 = 11+13 = 7 + 17 = P_5 + P_6 = P_4 + P_7$

(PW31) $22 = 3+19 = 5 + 17 = P_2 + P_8 = P_3 + P_7$

(PW32) $20 = 7+13 = 3 + 17 = P_4 + P_6 = P_2 + P_7$

(PW33) $18 = 5+13 = 7 + 11 = P_3 + P_6 = P_4 + P_5$

(PW34) $16 = 3+13 = 5 + 11 = P_2 + P_6 = P_3 + P_5$

(PW35) $14 = 7+7 = 3 + 11 = P_4 + P_4 = P_2 + P_5$

Thus, any even number greater than twelve (12) may be expressed as the sum of two primes in more than one way. Q.E.D.

Chapter 7

Proofs of Limited Prime Triplets & Singlets plus Proofs of Andrica's, Legendre's, and Brocard's Conjectures

It will now be shown that there can only be one sequence of prime triplets. Prime triplets are three primes whose successive difference is two (2). The sequence {3, 5, 7} is the first known sequence of prime triplets.

Consider the following six sequences which generate all the naturals in six (6) columns.

(PST1.1) $\{6I - 5 \mid I = 1, 2, 3, \ldots \infty\}$

Proofs of Limited Prime Triplets & Singlets plus
Proofs of Andrica's, Legendre's & Brocard's Conjecture

(PST1.2) $\{6J - 4, \mid J = 1, 2, 3, \ldots \infty\}$

(PST1.3) $\{6K - 3, \mid K = 1, 2, 3, \ldots \infty\}$

(PST1.4) $\{6L - 2, \mid L = 1, 2, 3, \ldots \infty\}$

(PST1.5) $\{6M - 1, \mid M = 1, 2, 3, \ldots \infty\}$

(PST1.6) $\{6N \mid N = 1, 2, 3, \ldots \infty\}$

Col 1	Col 2	Col 3	Col 4	Col 5	Col 6
1	2	3	4	5	6
7	8	9	10	11	12
13	14	15	16	17	18
19	20	21	22	23	24
25	26	27	28	29	30
31	32	33	34	35	36
37	38	39	40	41	42
43	44	45	46	47	48
49	50	51	52	53	54
55	56	57	58	59	etc.

Table 2

Triplet Proof

Equations (PST0.1) to (PST0.6) generates table 2. All prime numbers greater than 3 have the form of either

(PST2) $6(I+1) - 5 = 6I + 1$ or

(PST3) $6M - 1$

where I and M are natural indices. Suppose the first prime has the form (column 5)

(PST4) $P_K = 6M - 1$

then the next closest prime of a possible triplet must have the form (column 1)

(PST5) $P_{K+1} = 6I + 1$

The difference between these two primes is indeed two (2) since for the case where $I = M$

Proofs of Limited Prime Triplets & Singlets plus
Proofs of Andrica's, Legendre's & Brocard's Conjecture

(PST6) $P_{K+1} - P_K = 6I + 1 - (6I - 1) = 2$

However the next closest prime in the natural sequences of (PST1.1) to (PST1.6) must be

(PST7) $P_{K+2} = 6I - 1 + 6 = 6I + 5$

The difference between these last two primes is

(PST8) $P_{K+2} - P_{K+1} = 6I + 5 - (6I + 1) = 4$

which means that three primes (P_K, P_{K+1}, P_{K+2}) cannot exist where their successive differences are two (2) starting with a prime, P_K of form $6M - 1$.

If the first prime had been of the form (column 1)

(PST9) $P_L = 6I + 1$

then the next closest (I=M) prime of a possible triplet has the form $(6(M+1) - 1) = 6M+5$, so

(PST10) $P_{L+1} = 6M + 5 = 6I + 5$

The difference between these two primes is

(PST11) $P_{L+1} - P_L = 6I + 5 - (6I + 1) = 4$

which cannot be part of a triplet. Therefore, neither the three primes (P_K, P_{K+1}, P_{K+2}) or (P_L, P_{L+1}, P_{L+2}) cannot exist where their successive differences are two (2). Q.E.D.

Singlet Proof

This proof is only presented for completeness. A set of singlet primes is two primes which differ by one (1). The example is

(PST12) $P_1 = 2$

and

(PST13) $P_2 = 3$

The proof that these two primes form the only singlets is the fact that all primes > 3 follow the form of equations

(PST2) 6M − 1 or

(PST3) 6I + 1

Therefore, the minimum difference N_D, between any two primes (greater than three(3)) is when I=M to yield

(PST14) $N_D = 6I + 1 − (6I − 1) = 2 > 1$

Thus, there can only be one set of prime singlets.
Q. E. D.

Proof of Andrica's Conjecture

Andrica's conjecture states that the square root of the (n + 1)th prime less the square root of the nth prime is always less than unity. In other words,

(PST15) $\sqrt{P_{n+1}} - \sqrt{P_n} < 1$

Let us assume this conjecture is false. This means

(PST15.1) $\sqrt{P_{n+1}} - \sqrt{P_n} \geq 1$

If we multiply both sides of this inequality by the quantity $(\sqrt{P_{n+1}} + \sqrt{P_n})$ then (PST15.1) becomes

(PST16) $(\sqrt{P_{n+1}} + \sqrt{P_n})(\sqrt{P_{n+1}} - \sqrt{P_n}) \geq (\sqrt{P_{n+1}} + \sqrt{P_n})$

which simplifies to

(PST17) $P_{n+1} - P_n \geq \sqrt{P_{n+1}} + \sqrt{P_n}$

Note that the right hand side of relation (PST17) monotonically increases as n gets large. However, the left hand side oscillates, which rules out any possibility of equality. In fact, the left hand side has the value of 2 for all twin primes and for which there are no two consecutive primes which satisfy (PST17). Thus, the assumption that Andrica's conjecture is false, has logically produced a contradiction for every set of twin primes. Therefore, since relation (PST17) has been shown to be false, Andrica's conjecture of equation

(PST15) $\sqrt{P_{n+1}} - \sqrt{P_n} < 1$

must be true. Q.E.D.

Proof of Legendre's Conjecture

Legendre's conjecture says that for any natural number, n there exists a prime P, between n^2 and $(n + 1)^2$. In other words

(PST18) $n^2 < P < (n + 1)^2$

Let us assume that Legendre's conjecture is false, which means that for some k,

(PST18.1) $P_k < n^2$

and the next prime

(PST18.2) $P_{k+1} > (n + 1)^2$

Taking square roots and rewriting inequalities (PST18.1) and (PST18.2), we have

(PST18.3) $\sqrt{P_k} < n$
and

(PST18.4) $\sqrt{P_{k+1}} > n + 1$

Negating equation (PST18.3) we have

(PST18.5) $-\sqrt{P_k} > -n$

Adding inequalities (PST18.4) and (PST18.5) we get

(PST18.6) $\sqrt{P_{k+1}} - \sqrt{P_k} > 1$

which cannot be true by the proof of Andrica's conjecture above. Thus, Legendre's conjecture that

(PST18) $n^2 < P < (n + 1)^2$

must be true. Q.E.D.

Proof of Brocard's Conjecture

Recall the Rosser - Schoenfeld (RS) theorem renumbered for this section is

(PST20.0) $N/Ln(N) < \pi(N) < 1.25506 N/Ln(N)$

where $N > 16$. Brocard's conjecture states that

(PST20.1) $\pi(P_{n+1}^2) - \pi(P_n^2) \geq 4$

for all $n > 1$

Utilizing the left hand side of the RS theorem,

(PST20.2) $\pi(P_{n+1}^2) > (P_{n+1}^2)/Ln(P_{n+1}^2)$

and

(PST20.3) $\pi(P_n^2) > (P_n^2)/Ln(P_n^2)$

which means that

(PST20.4) $\pi(P_{n+1}^2) - \pi(P_n^2) >$
$\qquad P_{n+1}^2/Ln(P_{n+1}^2) - P_n^2/Ln(P_n^2)$

Recall the integer part of a real number R is denoted by $\lessdot R \gtrdot$ and since

(PST20.5) $\lessdot P_{n+1}^2/Ln(P_{n+1}^2) - P_n^2/Ln(P_n^2) \gtrdot <$
$\qquad P_{n+1}^2/Ln(P_{n+1}^2) - P_n^2/Ln(P_n^2)$

it follows that

(PST20.6) $\pi(P_{n+1}^2) - \pi(P_n^2) >$
$\qquad \lessdot P_{n+1}^2/Ln(P_{n+1}^2) - P_n^2/Ln(P_n^2) \gtrdot$

The right hand side of relation (PST20.6) is a minimum when n = 2 or when $P_{n+1} = P_3 = 5$ and $P_n = P_2 = 3$. For this case

(PST20.7) $\lessgtr P_3^2/Ln(P_3^2) - P_2^2/Ln(P_2^2) \gtrless = 3$

The left hand side of relation (PST20.6) for this case is

(PST20.8) $\pi(P_3^2) - \pi(P_2^2) = 5$

since there are four (4) primes before $P_2^2 = 9$ and nine (9) primes before $P_3^2 = 25$.

The right hand side of relation (PST20.6) when n = 3 or when $P_{n+1} = P_4 = 7$ and $P_n = P_3 = 5$. For this case

(PST20.7) $\lessgtr P_4^2/Ln(P_3^2) - P_3^2/Ln(P_2^2) \gtrless = 4$

For all other cases, when n > 3 the right hand side of relation (PST20.6) is greater than 4. Thus, Brocard's conjecture that

(PST20.1) $\pi(P_{n+1}^2) - \pi(P_n^2) \geq 4$

for n > 1, by both the prime number theorem (PNT) and the Rosser - Schoenfeld (RS) theorem must be true. Q.E.D.

Chapter 8

Proofs of Twin Prime Conjecture and Any Prime > 3 is the Average of Two Other Odd Primes

It will now be demonstrated (via the PNT) that there exists an infinite number of twin primes (two primes that differ by two (2)). After, this, it will also be shown that a given prime greater than three (3) must be an average of two other odd primes. Lastly, a proof by construction (with help from the proof of the binary conjecture) of the twin prime conjecture is presented.

Twin Prime Preliminaries

Since all the naturals can be expressed in six (6) columns (Refer to table 2) and the fact that all primes > 3 have the either the form of an alpha prime (column 1) which is

(PST2) $6I + 1$

or the form of an epsilon prime (column 5) which is

(PST3) $6M - 1$

where $I > 0$ and $M > 0$ are both natural indices. Thus, the only possibility of twin primes is that some epsilon prime must be subtracted from some alpha prime to yield the number two (2). The only exception to this is the first set of twins ($P_2 = 3$, $P_3 = 5$) since P_2 is neither an alpha or epsilon prime. If any alpha prime is subtracted from a larger epsilon prime, the minimal difference is always greater than two (2) and so the pair cannot be twins. Referring to table 2, column 6 elements (6N) minus column 1 elements (6I + 1) (containing all alpha primes) yield column 5 elements (containing all epsilon primes) (6M − 1). Thus,

(PUT0) $6N − (6I − 5) = 6(N − I) + 5 = 6M − 1$

where $N − I = M − 1$. Similarly, subtracting a column 5 element from a column 6 element results in a column 1 element.

(PUT0.1) $6N − (6M − 1) = 6(N − M) + 1 = 6I + 1$

again, where $N − M = I$. Note also that subtracting a column 1 element from a column 2 element results in another column 1 element.

(PUT0.2) $6J − 4 − (6I_1 − 5) = 6(J − I_1) + 1 = 6I − 5$

again, where $J − I_1 = I − 1$.

Demonstration of Twin Prime Conjecture via PNT

By the prime number theorem as expressed by

(T1) $\pi(N) \approx N/Ln(N)$

the probability, $\Psi(N)$ that N is a prime is

(T4) $\Psi(N) \approx 1/lnN$

By the PNT, the probability that $6I - 5 = 6N + 1$ for $I = N+1$ is an alpha prime is

(PUT1) $\Psi_{6N+1} \approx 1/Ln(6N+1)$

and the probability that $6M-1 = 6N - 1$ for $M = N$ is an epsilon prime is

(PUT2) $\Psi_{6N-1} \approx 1/Ln(6N-1)$

thus, the probability that 6N+1 and 6N−1 are both primes is the joint probability of twin primes or

(PUT3) $\Psi_T \approx \Psi_{6N+1}\Psi_{6N-1} \approx 1/[Ln(6N+1)Ln(6N-1)]$

so in a run of naturals from 1 to 6N+2, a minimum number of twins is

(PUT4) $N_T = \langle (6N+2)/(Ln(6N+1)Ln(6N-1)) \rangle$

This estimate is low since it is known that the actual number of primes is greater than the PNT estimate

(T2) $N/Ln(N) < \pi(N)$

so define a number less than N_T and call it

(PUT5) $N_L = \langle (6N+2)/Ln^2(6N+1) \rangle < N_T$

If we take the limit of N_L as N approaches infinity, by a double application of L'Hopital's rule, we have

60 Primal Proofs

(PUT6) $\lim_{N \to \infty} N_L = \lim_{N \to \infty} \prec(6N+2)/Ln^2(6N+1)\succ$

$= \lim_{N \to \infty} \prec(d(6N+2)/dN)/(d(Ln^2(6N+1)/dN)\succ$

$= \lim_{N \to \infty} \prec(6N+1)/2Ln(6N+1)\succ$

$= \text{Lim}_{N \to \infty} \prec d(6N+1)/dN)/2d(Ln(6N+1))/dN)\succ$

$= \text{Lim}_{N \to \infty} \prec(6N+1)/2\succ = \infty$

implying

(PUT7) $\lim_{N \to \infty} N_L = \infty$

thus, a minimal number of twins in equation

(PUT4) $N_T = \prec(6N+2)/(Ln(6N+1)Ln(6N-1))\succ$

being more than N_L by equation

(PUT5) $N_L = \prec(6N+2)/Ln^2(6N+1)\succ < N_T$

means that

(PUT8) $\lim_{N \to \infty} N_T = \infty$

demonstrating and implying an infinite number of twin primes.

Proof Prime > 3 is the Average of Two Other Odd Primes

In chapter 4, it was demonstrated that by the PNT, any even number greater than or equal to twenty four (24) satisfies Goldbach's binary conjecture. Equations

(AB2) $N_{OP} \approx \prec(E_J - Ln(E_J))/(Ln^2(E_J))\succ$

and

(AB5) $N_W = \prec N_{OP}/2 \succ$

express that N_{OP} is the number of odd primes resulting from subtracting all odd primes less than some even number, E_J from that even number. N_W is a minimal number of ways that two primes add up to E_J. Note that N_W was greater than zero (other than the case of $E_1 = 2$) for E_J greater than or equal to twenty-four (24). It also turns out that for any even number, E_J greater than or equal to eighty-four (84), there are at least two ways in which two odd primes add up to the even number. This means that there exists another case other than when $E_J = 2P_{K+2}$ where $K > 0$. Thus, any prime greater than forty-one (41) when doubled, must be the sum of two other odd primes or

(PAP1) $2P_{C+2} = P_{A+1} + P_{B+2}$

where A, B, or C > 0 and when rewritten is

(PAP2) $P_{C+2} = (P_{A+1} + P_{B+2})/2$

where C > 11 since $P_{13} = 41$. For cases of C ≤ 11, observe that

(PAP3) $P_{13} = 41 = (3 + 79)/2 = (P_2 + P_{22})/2$

(PAP4) $P_{12} = 37 = (3 + 71)/2 = (P_2 + P_{20})/2$

(PAP5) $P_{11} = 31 = (3 + 59)/2 = (P_2 + P_{17})/2$

(PAP6) $P_{10} = 29 = (5 + 53)/2 = (P_3 + P_{16})/2$

(PAP7) $P_9 = 23 = (3 + 43)/2 = (P_2 + P_{14})/2$

(PAP8) $P_8 = 19 = (7 + 31)/2 = (P_4 + P_{11})/2$

(PAP9) $P_7 = 17 = (3 + 31)/2 = (P_2 + P_{11})/2$

(PAP10) $P_6 = 13 = (3 + 23)/2 = (P_2 + P_9)/2$

(PAP11) $P_5 = 11 = (3 + 19)/2 = (P_2 + P_8)/2$

(PAP12) $P_4 = 7 = (3 + 11)/2 = (P_2 + P_5)/2$

(PAP13) $P_3 = 5 = (3 + 7)/2 = (P_2 + P_4)/2$

64 Primal Proofs

thus, any prime greater than 3 is the average of two other odd primes. Q.E.D.

Proof of Twin Prime Conjecture by Construction

In the twin prime preliminary section of this chapter, it has been shown that twin primes greater than (3, 5) can only occur when the difference between a column 1 (refer to table 2) (alpha) prime and column 5 (epsilon) primes is two (2). Moreover, it was demonstrated that column 6 evens are the sum of column 1 odds and column 5 odds and column 2 evens are the sum of two column 1 elements.

By the proof of the binary conjecture, for any given $N \geq 3$

(PTW1) $E_{3N+1} = 6N + 2 = P(\alpha, A_{3N+1}) + P(\alpha, B_{3N+1})$

and

(PTW2) $E_{3N} = 6N = P(\alpha, A_{3N}) + P(\varepsilon, B_{3N})$

and by the fact that there is a plurality of ways in which an even number greater than twelve (12) is the sum of two primes (see chapter 6) we also can have

(PTW3) $E_{3N+1} = 6N + 2 = P(\alpha, C_{3N+1}) + P(\alpha, D_{3N+1})$

and

(PTW4) $E_{3N} = 6N = P(\alpha, C_{3N}) + P(\varepsilon, D_{3N})$

where $P(\alpha, A_{3N+1})$, $P(\alpha, B_{3N+1})$, $P(\alpha, C_{3N+1})$, $P(\alpha, D_{3N+1})$, are four alpha primes belonging to (referring to table 1) E_{3N+1}'s row. $P(\alpha, A_{3N})$, $P(\varepsilon, B_{3N})$, $P(\alpha, C_{3N})$, $P(\varepsilon, D_{3N})$, are four more primes two (alpha and epsilon primes) belonging to (referring to table 1) E_{3N}'s row. All eight primes must be less than 6N+2. Subtracting equations (PTW2) from (PTW1) yields

(PTW5) $E_{3N+1} - E_{3N} = 2 =$
$P(\alpha, A_{3N+1}) - P(\alpha, A_{3N}) + P(\alpha, B_{3N+1}) - P(\varepsilon, B_{3N})$

Subtracting equations (PTW4) from (PTW3) yields

(PTW6) $E_{3N+1} - E_{3N} = 2 =$
$P(\alpha,C_{3N+1}) - P(\alpha,C_{3N}) + P(\alpha,D_{3N+1}) - P(\varepsilon,D_{3N})$

Referring to table 1, $P(\alpha,A_{3N+1})$ and $P(\alpha,C_{3N+1})$ must occur in columns (say columns O_{JK} and O_{JM}) to the right of column O_{J1} and belonging to E_{3N+1}'s row. However, since all columns are a natural run of odds, both $P(\alpha,A_{3N+1})$ and $P(\alpha,C_{3N+1})$ must also occur in previous columns O_{JK-1}, and O_{JM-1} belonging to E_{3N}'s row. These are designated by $P(\alpha,A_{3N})$ and $P(\alpha,C_{3N})$. Thus, there must be some $P(\alpha,A_{3N}) = P(\alpha,A_{3N+1})$ and some $P(\alpha,C_{3N}) = P(\alpha,C_{3N+1})$. Hence, equations (PTW5) and (PTW6) reduce to

(PTW7) $P(\alpha,B_{3N+1}) - P(\varepsilon,B_{3N}) = 2$

and

(PTW8) $P(\alpha,D_{3N+1}) - P(\varepsilon,D_{3N}) = 2$

describing two sets of twins. Depending on how many ways two primes can add up to the even numbers designated by 6N + 2 and 6N, there could be more equations such as (PTW7) and (PTW8) describing more twins per set of consecutive (6N+2 and 6N) evens. Choosing the highest numerical set of twins for a given set of consecutive even numbers of the form (6N+2 and 6N) would eliminate twin duplicates for increasing N. This means that since N is unlimited, there must be an unlimited set of unique twin primes Q.E.D.

Chapter 9

Proofs of the Riemann Hypothesis and Unlimited Primes

A proof of the Riemann hypothesis (that the real part of all non-trivial zeros of the zeta function all lie on a line in the complex plane, parallel to the y axis and designated by $x = 1/2$) by construction will be offered first. Next a proof of the Riemann hypothesis by contradiction is presented. Then, Euclid's proof of unlimited primes (for completeness) by contradiction will be presented. Finally, an alternate

PNT proof of unlimited primes by construction is presented.

Proof of the Riemann Hypothesis by Construction

The starting point for this proof is Riemann's extended zeta function given by equation (T5) and renumbered for this section as

(R1.0) $\zeta(s) = 2^s \pi^{s-1} \sin(\pi s/2) \Gamma(1-s) \zeta(1-s)$

where Γ is the gamma function and Sin is the normal trigonometric sine function. Here π is the well known constant defined to be the circumference of a circle divided by its diameter. $\zeta(s)$ is the extended zeta function of s, where s is assumed to be a complex number not equal to one ($s \neq 1$). Thus,

(R1.1) $s = x + iy$

then by (T8) $A^{X+iY} = A^X \{\cos(\ln A^Y) + i\sin(\ln A^Y)\}$

(R2.0) $2^s = 2^x \{\text{Cos}[\text{Ln}(2^y)] + i\text{Sin}[\text{Ln}(2^y)]\}$

and by inspection, cannot be equal to zero unless $x = -\infty$. Similarly,

(R2.1) $\pi^{s-1} = \pi^{x-1} \{\text{Cos}[\text{Ln}(\pi^y)] + i\text{Sin}[\text{Ln}(\pi^y)]\}$

cannot be zero unless $x = -\infty$. Moreover,

(R2.2) $\text{Sin}(\pi s/2) = \text{Sin}(\pi x/2)\text{Cosh}(\pi y/2) + i\text{Cos}(\pi x/2)\text{Sinh}(\pi y/2)$

where Cosh and Sinh are the normal hyperbolic sine and hyperbolic cosine functions. Note the only way $\text{Sin}(\pi s/2)$ can be zero is on the x axis in which case $y = 0$. It is also known that

(R2.3) $\Gamma(1-s) \neq 0$

for $\text{Re}(1-s) > 0$

Rewriting equation

(R1.0) $\zeta(s) = 2^s \pi^{s-1} \operatorname{Sin}(\pi s/2) \Gamma(1-s) \zeta(1-s)$

and setting the condition for zeros yields

(R3.0) $\zeta(s_0) = 2^{s_0} \pi^{s_0-1} \operatorname{Sin}(\pi s_0/2) \Gamma(1-s_0) \zeta(1-s_0) = 0$

where by equation (R1.1) $s = x + iy$, implies

(R1.2) $s_0 = x_0 + iy_0$

where s_0 is a general zero of the zeta function. Restricting all the non-trivial zeros to not lie on the x axis means

(R1.3) $y_0 \neq 0$

forcing s_0 to be a general non-trivial zero. This means that equation (R2.0) becomes

(R2.0.1) $2^{s_0} = 2^{x_0}\{\operatorname{Cos}[\operatorname{Ln}(2^{y_0})] + i\operatorname{Sin}[\operatorname{Ln}(2^{y_0})]\}$

$\neq 0$ assuming $x_0 \neq -\infty$

72 Primal Proofs

and equation (R2.1) becomes

(R2.1.1) $\pi^{s_0-1} = \pi^{x_0-1}\{Cos[Ln(\pi^{y_0})] +$

$iSin[Ln(\pi^{y_0})]\}$

$\neq 0$ assuming $x_0 \neq -\infty$

and equation (R2.2) becomes

(R2.2.1) $Sin(\pi s_0/2) = Sin(\pi x_0/2)Cosh(\pi y_0/2) +$

$iCos(\pi x_0/2)Sinh(\pi y_0/2)$

$\neq 0$ assuming $y_0 \neq 0$

and equation (R2.3) becomes

(R2.3.1) $\Gamma(1-s_0) \neq 0$

for $Re(1-s_0) > 0$

Dividing both sides of equation (R3.0) by the non-zero product $2^{s_0}\pi^{s_0-1}Sin(\pi s_0/2)\Gamma(1-s_0)$ (because of equations R2.0.1, R2.1.1, R2.2.1 and R2.3.1) yields

(R3.1) $\zeta(s_0) = \zeta(1-s_0) = 0$

However, it is known that the Riemann zeta function obeys

(R4.0) $\zeta(s) = [\zeta(s^*)]^* = \zeta^*(s^*)$

where s^* is the complex conjugate of s and ζ^* is the complex conjugate of ζ. Observe that since $\zeta(s_0) = 0$, then $\zeta^*(s_0^*) = 0^* = 0 = \zeta(s_0^*)$ which means

(R4.1) $\zeta(s_0) = \zeta(s_0^*) = 0$

and says non-trivial zeros of the zeta function can only occur in conjugate pairs symmetric about the real axis. Note also that if $x_0 > 0$ and $y_0 > 0$, then s_0 and $1-s_0^*$ are in the first quadrant of the complex plane. Moreover both s_0^* and $1-s_0$ are in the fourth quadrant of the complex plane.

If by (R3.1) that $\zeta(s_0) = \zeta(1-s_0) = 0$ and by (R4.1) that $\zeta(s_0) = \zeta(s_0^*) = 0$, it follows that

(R4.2) $\zeta(s_0^*) = \zeta(1-s_0) = 0$ and

74 Primal Proofs

(R4.2.1) $\zeta(s_0) = \zeta(1-s_0^*) = 0$

Let's now write Reimann's zeta function as its Laurent series (showing it in terms of its explicit argument) presented in the tools chapter (chapter 2) and renumbered for this section as

(R5) $\zeta(s) = 1/(s-1) + \gamma_0 + \gamma_1(s-1) + \gamma_2(s-1)^2 + \ldots +$

where the γ_k ($k = 0, 1, 2 \ldots$) are constants. Equation (R5) in terms of a general non trivial zero ($s_0 = x_0 + iy_0$) is

(R5.1) $\zeta(s_0) = 1/(s_0 - 1) + \gamma_0 + \gamma_1(s_0 - 1) +$
$\gamma_2(s_0 - 1)^2 + \ldots + = 0$

Equation (R5) written in terms of $(1-s_0^*)$ becomes

(R5.2) $\zeta(1-s_0^*) = -1/s_0^* + \gamma_0 - \gamma_1 s_0^* + \gamma_2 s_0^{*2} +$
$\ldots + \qquad = 0$

where both $s_0^* \neq 0$ and $s_0 \neq 1$. Note that (R5.1) and (R5.2) are both valid in the critical strip defined as $0 < x < 1$ where all non trivial zeros of the zeta function are known to exist.

However, the series of equation (R5.1) must be equivalent to the series of equation (R5.2) by equation (R4.2.1) $\zeta(s_0) = \zeta(1- s_0^*) = 0$. Moreover, since the Laurent series is unique, the only possibility is for a term by term equivalence, hence

(R6.0) $1 - s_0^* = s_0$

which means

(R6.1) $1 - (x_0 - iy_0) = x_0 + iy_0$

whose only solution is

(R6.2) $x_0 = 1/2$

Thus, the argument of the zeta function when the zeta function is a non-trivial zero, must have real part equal to one half. Q.E.D.

Proof of the Riemann Hypothesis by Contradiction

Assume that very far up ($y = y_{00}$) the critical strip, a pair of zeros ($s_{01} = x_{01} + iy_{00}$ and $s_{02} = x_{02} + iy_{00}$ not on the critical line) of the zeta ($\zeta(s)$) function occurs. These two zeros must be symmetric about the critical line, and since the zeta function is continuous between these zeros, it follows that the slope of the zeta function with respect to the real axis must be zero ($d\zeta(s)/dx = 0$) somewhere between these two points. Therefore,

(R7.0) $d\zeta(s)/dx = (d\zeta(s)/ds)(ds/dx) = 0$

Note that if

(R7.1) $d\zeta(s)/ds = 0$ and/or

(R7.2) $ds/dx = 0$

are true, then (R7.0) will be satisfied. Consider the case of (R7.1) $d\zeta(s)/ds = 0$. The solution of (R7.1) is

(R7.1.1) $\zeta(s) = C_1 + iC_2$

where both C_1 and C_2 are constants. However for $\zeta(s)$ to be zero (at s_{01} and s_{02} by assumption), both C_1 and C_2 must be zero. This is absurd since $\zeta(s)$ would have to be zero for all values of s.

The only other possibility that (R7.0) $d\zeta(s)/dx = (d\zeta(s)/ds)(ds/dx) = 0$ is for (R7.2) $ds/dx = 0$. This means that since

(R7.2.1) $s = x + iy$

(R7.2.1) $ds/dx = 1 + idy/dx = 0$

which implys

(R7.2.3) $dy/dx = -1/i = -i/i^2 = i$

whose solution is

(R7.2.4) $y = xi + K$

where K cannot be a function of x. Plugging (R7.2.4) into (R7.2.1) $s = x + iy$ results in

(R8.1) $s = x + i(xi + K) = iK$

Thus, (R7.0) $d\zeta(s)/dx = (d\zeta(s)/ds)(ds/dx) = 0$ can only be true for values of s on the y axis by (R8.1). This contradicts that $d\zeta(s)/dx$ must be zero somewhere within the critical strip between both assumed zeros (s_{01} and s_{02}) of zeta. This means that there cannot be symmetric zeros about the critical line within the critical strip. However, all non trivial zeros of the zeta function exist only in the critical strip. Therefore, all zeros of zeta must lie on the critical line where $x_0 = 1/2$. Q.E.D. Note that equation

(R3.1) $\zeta(s_0) = \zeta(1-s_0) = 0$ reduces to

(R3.2) $\zeta(1/2 + iy_0) = \zeta([1-(1/2 + iy_0)]) = 0$

and can be rewritten as

(R3.3) $\zeta(1/2 + iy_0) = \zeta(1/2 - iy_0) = 0$

which is equivalent to and verifies equation

(R4.1) $\zeta(s_0) = \zeta(s_0{}^*) = 0$

In summary, the only non-trivial zeros of the zeta function must have the form $s_0{}^\pm = 1/2 \pm iy_0$ where $y_0 \neq 0$. This means all zeta function non-trivial zeros lie on the line ($x_0 = 1/2$) and come as symmetric conjugate pairs about the real axis.

Euclid's Proof of Unlimited Primes

Euclid first showed that the number of primes is infinite by assuming that there is one last prime P_L in the natural number sequence with no other primes following P_L. He then expressed a number

N_L formed by adding one (1) to the product of all the primes as

(AP1) $N_L = (P_1)(P_2)(P_3)(P_4)...(P_L) + 1$

By assumption N_L must be composite. However, if N_L is divided by any of these primes, a remainder of one (1) will always result. This can be expressed by

(AP2) $Mod(N_L, P_J) = 1$

where P_J is taken from $\{P_1, P_2, P_3 ... P_L\}$. Thus, N_L does not contain any prime factors which means it must be a prime greater than P_L. This contradiction means that there is no last prime P_L in the natural number sequence. If there is no last prime, then they must go on forever. Q.E.D.

Unlimited Primes via PNT

The prime number theorem can also be used to prove unlimited primes by a single application of L'Hopital's rule. The number of primes less than or equal to some natural N by the prime number theorem is

(T1) $\pi(N) \approx N/Ln(N)$

Define

(AP1) $N_P = \langle N/Ln(N+1) \rangle$

and note relationships

(AP2) $N_P < N/Ln(N)$

and

(T2) $N/Ln(N) < \pi(N)$

which means that

(AP2.1) $N_P < N/Ln(N) < \pi(N)$

Letting N approach infinity we have

(AP3) $\lim_{N \to \infty} N_P = \lim_{N \to \infty} \prec(N)/Ln(N+1)\succ$

$\qquad = \lim_{N \to \infty} \prec(d(N)/dN)/(d(Ln(N+1)/dN)\succ$

$\qquad = \lim_{N \to \infty} \prec(N+1)\succ = \infty$

since by equations (AP2) and (T2) that

(AP4) $N_P < \pi(N)$

it follows that

(AP5) $\lim_{N \to \infty} \pi(N) = \infty$

thus, by the Prime number theorem (PNT) there is an infinite number of primes. Q.E.D.

Chapter 10

Prime Entanglement

All prime numbers greater than three (3) fall into two broad entangled forms. These have been called alpha (α) primes and epsilon (ε) primes. Alpha primes are of the form 6I + 1 and epsilon primes are of the form 6M − 1 where I and M are natural number indices. Another way of looking at this is to note that

(E0.0) Mod (P(α,j), 6) = 1

and

(E0.1) Mod $(P(\varepsilon,i), 6) = 5$

where $P(\alpha,j)$ and $P(\varepsilon,i)$ are the jth alpha prime and the ith epsilon prime respectively.

If all the natural numbers are listed in six columns, each column is specified by the sequence equations (PST1.1) to (PST1.6). These sequences

Col 1	Col 2	Col 3	Col 4	Col 5	Col 6
1	2	3	4	5	6
7	8	9	10	11	12
13	14	15	16	17	18
19	20	21	22	23	24
25	26	27	28	29	30
31	32	33	34	35	36
37	38	39	40	41	42
43	44	45	46	47	48
49	50	51	52	53	54
55	56	57	58	59	etc.

Table 2

generate table 2 which is reproduced above for convenience.

Proof All Primes Greater than 3 Are In Columns 1 and 5 of Table 2

Columns 2, 4 and 6 contain all the even numbers and thus cannot contain any odd primes. Column 3 only contains the prime three (3) and multiples of three (3). This implies that the rest of all the odd primes are in columns 1 and 5.

Note that columns 1, 3 and 5 contain all the odd naturals. Formulas for the sequences represented by any of these columns excluding the first three elements of row one (1, 2, 3) are as follows:

The sequence of column 1 is

(E3.1) $\{O(a,i)\} = \{6i + 1 \mid i = 1, 2, 3, .. \infty\}$

The sequence of column 2 is

(E3.2) $\{E(b,i)\} = \{6i + 2 \mid i = 1, 2, 3, .. \infty\}$

The sequence of column 3 is

(E3.3) $\{O(c,i)\} = \{6i + 3 \mid i = 1, 2, 3, .. \infty\}$

The sequence of column 4 is

(E3.4) $\{E(d,i)\} = \{6i - 2 \mid i = 1, 2, 3, .. \infty\}$

The sequence of column 5 is

(E3.5) $\{O(e,i)\} = \{6i - 1 \mid i = 1, 2, 3, .. \infty\}$

The sequence of column 6 is

(E3.6) $\{E(f,i)\} = \{6i \mid i = 1, 2, 3, .. \infty\}$

The first argument of any column element has been labeled as a, b, c, d, e, f. which respectively denotes the 1st, 2nd, 3rd, 4rth, 5th, and 6th column. The second (i) argument of any column element denotes the ith element of that column. Recall that O and E refer to odd and even naturals respectively.

Prime Entanglement 87

The column 5 sequence may also be represented by substituting i = n + 1, in which case equation (E3.5) becomes

(E3.5.1) $\{O(e, n + 1)\} = \{6n + 5 \mid n = 0, 1, 2, .. \infty\}$

By Direchlet's theorem (see glossary) both 6i + 1 and 6n + 5 generate an infinite number of primes as i and n take on an infinite number of values. Therefore, all primes > 3 must be in columns 1 and 5. Q.E.D.

Symbolically, for certain values, i_α and j_ε

(E1.0) $P(\alpha, k) = 6j_\alpha + 1$

and

(E2.0) $P(\varepsilon, m) = 6i_\varepsilon - 1$

where j_α and i_ε are the indices into the kth alpha prime and mth epsilon prime respectively. Note that $P(\varepsilon, 1) = 5$ and $P(\alpha, 1) = 7$.

Selective Binary Relationships

Subtracting a column 1 element from a column 2 element results in

(E5.0) $E(b,k) - O(a,j) = 6k+2 - (6j+1) = 6(k-j) + 1$

Letting $m = k - j$, reduces (E5.0) to

(E5.1) $E(b,k) - O(a,j) = 6m + 1 = O(a,m)$

which means subtracting a column 1 element from a column 2 element produces another column 1 element.

Subtracting a column 5 element from a column 4 element results in

(E6.0) $E(d,k) - O(e,j) = 6k-2 - (6j-1) = 6(k-j) - 1$

Letting $m = k - j$, reduces (E12.0) to

(E6.1) $E(d,k) - O(e,j) = 6m - 1 = O(e,m)$

which means subtracting a column 5 element from a column 4 element produces another column 5 element.

Subtracting a column 5 element from a column 2 element results in

(E7.0) $E(b,k) - O(e,j) = 6k - 6j + 3 = 6(k-j) + 3$

Letting $m = k - j$, reduces (E7.0) to

(E7.1) $E(b,k) - O(e,j) = 6m + 3 = O(c,m)$

which means subtracting a column 5 element from a column 2 element produces a column 3 element.

Subtracting a column 2 element from a column 5 element results in

(E8.0) $O(e,k) - E(b,j) = 6k - 6j - 3 = 6(k-j) - 3$

Letting $m = k - j - 1$, reduces (E8.0) to

(E8.1) $O(e,k) - E(b,j) = 6m + 3 = O(c,m)$

which means subtracting a column 2 element from a column 5 element produces a column 3 element.

Thus, there are certain relationships between the 1st and 5th columns (which contain all primes > 3) that represent entanglement. Entanglement, loosely defined means that certain knowledge of the values of column 1 elements automatically imply certain knowledge about values of column 5 elements.

Column 1 contains all the alpha primes and it also contains all the alpha type odd composites. Column 5 contains all the epsilon primes and all epsilon type odd composites. Hence, both column 1 primes and column 1 odd composites are entangled with column 5 primes and column 5 odd composites. The next chapter on odd composites will demonstrate definite entangled relationships between column 1 and column 5 odd composites.

Chapter 11

Odd Composites

Sequences for all the odd composites will now be developed. Please refer to the six (6) columns of table 2. Recall these columns were consecutively labeled as a, b, c, d, e f. Column 1 contains all the alpha primes and column 5 contains all the epsilon primes.

Column 3 Odd Composites

The sequence given by

(E3.3) $\{O(c,i)\} = \{6i + 3 \mid i = 1, 2, 3, .. \infty\}$

92 Primal Proofs

represents all odd composites that are divisible by the prime 3 (which is neither an alpha or epsilon prime). This can be shown by factoring the element of sequence (E3.3) as

(O1.0) $O(c,i) = 6i + 3 = 3(2i + 1) = \Theta(c,0,i)$

where the argument of the composite $\Theta(c,0,i)$ denotes the ith element of column 3 composites generated by the $j = 0$ index into the $P_{j+2} = P_2 = 3$ prime.

The notation of equation (O1.0) can be further refined to yield

(O1.1) $O(c,m) = 6m + 3 = 3[2m + 1] = \Theta(c,0,m)$

where the dummy index i has been replaced by m and where $\Theta(c,0,m)$ will be the adopted symbol for an odd composite in column 3. Note that the range of this index is $1 \leq m \leq \infty$. Thus, the sequence of column 3 odd composites is given by

(O1.2) $\{\Theta(c,0,m)\} = \{6m + 3 \mid m = 1, 2, 3, ..\infty\}$

Odd composites also exist in columns 1 and 5. Formulas for these sequences of odd composites will now be developed. In the foregoing analysis, please note that $1 \leq i, j, m, n \leq \infty$.

Note that $P_{j+2} > 3$, since the case where $P_2 = 3$ is given by the sequence of column 3 odd composites above. The column 3 odd composite index and the column 3 odd composite sequence is reproduced below for convenience.

(O1.1) $O(c,m) = 6m + 3 = 3[2m + 1] = \Theta(c,0,m)$

(O1.2) $\{\Theta(c,0,m)\} = \{6m + 3 \mid m = 1, 2, 3, ..\infty\}$

Column 1 Odd Composites

Odd composite elements in column 1 will take the form

(O2.0) $\Theta(a,j,i) = P_{j+2}(2i + 1) = 6m(a,j,i) + 1$

where the notation $m(a,j,i)$ means the mth index (which depends on the $(j+2)$th prime and the dummy index i) into the corresponding odd composite of column 1.

If $j = 1$, then $P_{j+2} = P_3 = 5$, equation (O2.0) reduces to the Diophantine equation

(O2.1) $5(2i + 1) = 6m(a,1,i) - 1$

which simplifies to

(O2.2) $3m(a,1,i) - 5i = 2$

Letting $i = 3n - 1$, equation (O2.2) becomes

(O2.3) $m(a,1,n) = 5n - 1$

which is the nth index into the column 1 odd composites (for $P_3 = 5$) given by

(O2.4) $\Theta(a,1,n) = 6m(a,1,n) + 1 = 30n - 5$

and producing the corresponding sequence (generated by $P_3 = 5$) of column 1 odd composites as

(O2.5) $\{\Theta(a,1,n)\} = (30n - 5 \mid n = 1, 2, 3, .. \infty\}$

If $j = 2$, then $P_{j+2} = P_4 = 7$, equation (O2.0) reduces to the Diophantine equation

(O2.6) $7(2i + 1) = 6m(a,2,i) - 1$

which simplifies to

(O2.7) $3m(a,2,i) - 7i = 3$

Letting $i = 3n$, and dividing by 3, equation (O2.7) becomes

(O2.8) $m(a,2,n) = 7n + 1$

which is the nth index into the column 1 odd composites (for $P_4 = 7$) given by

(O2.9) $\Theta(a,2,n) = 6m(a,2,n) + 1 = 42n + 7$

and producing the corresponding sequence (generated by $P_4 = 7$) of column 1 odd composites given by

(O2.10) $\{\Theta(a,1,n)\} = \{42n + 7 \mid n = 1, 2, 3, .. \infty\}$

Note that the solutions for the odd composites of column 1 for the primes 5 and 7 were achieved by either substituting the transformation

(O4.1) $i = 3n - 1$

when P_{j+2} is an epsilon prime (column 5) and

(O4.2) $i = 3n$

when P_{j+2} is an alpha prime (column 1). Note that

(O4.3) $\text{Mod}((P_{j+2}-1)/2, 3) = 0$

for alpha primes (primes in column 1) and

(O4.4) $\text{Mod}((P_{j+2}-1)/2, 3) = 2$

for epsilon primes (primes in column 5). Thus, the transformation

(O4.5) $i = 3n - \text{Mod}((P_{j+2} - 1)/2, 3)/2$

will simultaneously accomplish both equations

(O4.1) $i = 3n - 1$ and

(O4.2) $i = 3n$

Proceeding with the general solution of all column 1 composites, equation

(O2.0) $\Theta(a,j,i) = P_{j+2}(2i + 1) = 6m(a,j,i) + 1$

reduces to

(O5.0) $3m(a,j,i) = iP_{j+2} + (P_{j+2} - 1)/2$

Applying transformation (O4.5) to equation (O5.0) and solving for $m(a,j,n)$ yields

(O5.1) $m(a,j,n) = nP_{j+2} +$

$(P_{j+2}[1-\text{Mod}((P_{j+2}-1)/2, 3)] - 1)/6$

which is the index into all the column 1 odd composites. Thus, the element of all column 1 odd composites is

(O5.2) $\Theta(a,j,n) = 6m(a,j,n) + 1 =$

$6nP_{j+2} + P_{j+2}[1-\text{Mod}((P_{j+2}-1)/2, 3)]$

which generates the sequence of all the column 1 odd composites and given by

(O5.3) $\{\Theta(a,j,n)\} = \{6m(a,j,n) + 1 \mid j, n = 1, 2, .. \infty\}$

As a check, for $j = 1$, equation (O5.1) is

(O5.4) $m(a,1,n) = 5n + (5(-1) - 1)/6 = 5n - 1$

which is the same as

(O2.3) $m(a,1,n) = 5n - 1$

For $j = 2$, equation (O5.1) is

(O5.5) $m(a,2,n) = 7n + (7(1) - 1)/6 = 7n + 1$

which is the same as

(O2.8) $m(a,2,n) = 7n + 1$

Column 5 Odd Composites

Odd composites in column 5 take the form

(O3.0) $\Theta(e,j,i) = P_{j+2}(2i + 1) = 6m(e,j,i) - 1$

If $j = 1$, then $P_{j+2} = P_3 = 5$, equation (O3.0) reduces to the Diophantine equation

(O6.0) $5(2i + 1) = 6m(e,1,i) - 1$

which simplifies to

(O6.1) $3m(e,1,i) - 5i = 3$

Letting $i = 3n$, equation (O6.1) becomes

(O6.2) $m(e,1,n) = 5n + 1$

which is the nth index into the column 5 odd composites (for $P_3 = 5$) given by

(O6.3) $\Theta(e,1,n) = 6m(e,1,n) - 1 = 30n + 5$

and producing the corresponding sequence (generated by $P_3 = 5$) of column 1 odd composites as

(O6.4) $\{\Theta(e,1,n)\} = \{30n + 5 \mid n = 1, 2, 3, .. \infty\}$

If $j = 2$, then $P_{j+2} = P_4 = 7$, equation (O3.0) reduces to the Diophantine equation

(O6.5) $7(2i + 1) = 6m(e,2,i) - 1$

which simplifies to

(O6.6) $3m(e,2,i) - 7i = 4$

Letting $i = 3n - 1$, equation (O6.6) becomes

(O6.7) $m(e,2,n) = 7n - 1$

which is the nth index into the column 5 odd composites (for $P_4 = 7$) given by

(O6.8) $\Theta(e,2,n) = 6m(e,2,n) - 1 = 42n - 7$

102 Primal Proofs

and producing the corresponding sequence (generated by $P_4 = 7$) of column 5 odd composites as

(O6.9) $\{\Theta(e,2,n)\} = \{42n - 7 \mid n = 1, 2, 3, .. \infty\}$

Note that the solutions for the odd composites of column 5 for the primes 5 and 7 were achieved by either substituting the transformation

(O4.1) $i = 3n - 1$

when P_{j+2} is an epsilon prime (column 5) and

(O4.2) $i = 3n$

when P_{j+2} is an alpha prime (column 1). Note that

(O7.0) $\text{Mod}((P_{j+2} +1)/2), 3) = 0$

for epsilon primes (column 5) and

(O7.1) $\text{Mod}((P_{j+2}+1)/2), 3) = 1$

for alpha primes (column 1). Thus, the transformation

(O7.2) $i = 3n - \text{Mod}((P_{j+2} + 1)/2), 3)$

will simultaneously accomplish both equations

(O4.1) $i = 3n - 1$ and

(O4.2) $i = 3n$

Proceeding with the general solution of all column 5 composites, equation

(O3.0) $\Theta(e,j,i) = P_{j+2}(2i + 1) = 6m(e,j,i) - 1$

reduces to

(O8.0) $3m(e,j,i) = iP_{j+2} + (P_{j+2} +1)/2$

104 Primal Proofs

Applying transformation (O7.2) to equation (O8.0) and solving for m(e,j,n) yields

(O8.1) $m(e,j,n) = nP_{j+2} +$

$(P_{j+2}[1-2\text{Mod}((P_{j+2}+1)/2, 3)] +1)/6$

which is the index into all the column 5 odd composites. Thus, the element of all column 5 odd composites is

(O8.2) $\Theta(e,j,n) = 6m(e,j,n) - 1 =$

$6nP_{j+2} + P_{j+2}[1-2\text{Mod}((P_{j+2}+1)/2, 3)]$

which generates the sequence of all the column 5 odd composites and given by

(O8.3) $\{\Theta(e,j,n)\} = \{6m(e,j,n) - 1 \mid j, n = 1, 2, .. \infty\}$

As a check, for $j = 1$, equation (O8.1) is

(O8.4) $m(e,1,n) = 5n + (5(1) + 1)/6 = 5n + 1$

which is the same as

(O6.2) m(e,1,n) = 5n + 1

For j = 2, equation (O8.1) is

(O8.5) m(e,2,n) = 7n + (7(–1) + 1)/6 = 7n – 1

which is the same as

(O6.7) m(e,2,n) = 7n – 1

The following table (table 3) gives column 1 odd composite indices, column 1 odd composite elements, column 5 odd composite indices and column 5 odd composite elements. These were all generated from the general formulas above for primes from $P_3 = 5$ to $P_{13} = 41$. The natural dummy index, n ranges incrementally from one (1) to infinity.

a-index	a-comps	e-index	e-comps
$5n - 1$	$30n - 5$	$5n + 1$	$30n + 5$
$7n + 1$	$42n + 7$	$7n - 1$	$42n - 7$
$11n - 2$	$66n - 11$	$11n + 2$	$66n + 11$
$13n + 2$	$78n + 13$	$13n - 2$	$78n - 13$
$17n - 3$	$102n - 17$	$17n + 3$	$102n + 17$
$19n + 3$	$114n + 19$	$19n - 3$	$114n - 19$
$23n - 4$	$138n - 23$	$23n + 4$	$138n + 23$
$29n - 5$	$174n - 29$	$29n + 5$	$174n + 29$
$31n + 5$	$186n + 31$	$31n - 5$	$186n - 31$
$37n + 6$	$222n + 37$	$37n - 6$	$222n - 37$
$41n - 7$	$246n - 41$	$41n + 7$	$246n + 41$

etc...

Table 3

Table 3 also demonstrates how knowledge of index and sequence formulas for the column 1 composites (a-comps column of table 3) immediately gives rise to knowledge of index and sequence formulas for the columns 5 composites (e-comps column of table 3). Observe that changing the sign of the a-index and a-comps formulas generate the e-index and e-comps formulas! This is an example of the entanglement of the column 1 and column 5 odd composites as previously mentioned in Chapter 10 on prime entanglement. Note that if column 1 composites and column 5 composites are entangled, then so are the alpha primes entangled with the epsilon primes.

Chapter 12

Prime Number Generator

Two formulas which will generate all prime numbers greater than three (3) will be developed after a brief description of an ancient algorithm called the sieve of Eratosthenes (SOE) that produce primes. Eratosthenes was an ancient Greek mathematician.

The Sieve of Eratosthenes

The following algorithm is as follows. List the sequential natural numbers from the first prime (P_1 = 2) to the greatest natural number, N_G that will be checked for primality. Call this list the test list, L_T. Compute N_L as $N_L = \prec N_G^{1/2} \succ + 1$. N_L will mark the last number in L_T to be checked in the (BEGIN) section below. Put $P_1 = 2$ in the list of primes called L_P. Cross out 2 and all multiples of 2 in L_T. **(BEGIN)** The first number in the L_T list that is not crossed out is a prime. If this prime is greater than or equal to N_L, then go to the (FINISH) section. If this prime is less than N_L, then put this prime in L_P. Cross out this prime and all multiples of this prime from the test list L_T and go back to (BEGIN).

(FINISH) all the remaining numbers in L_T that are not crossed out are prime numbers and are put into the L_P list. No more numbers need to be crossed out. All primes in the test list L_T have been found and are now in the prime list L_P.

Note that the sieve of Eratosthenes (SOE) assumes that composites can be identified and systematically crossed out in the test list. Thus, if all the odd composites can be a-priori identified, then the SOE can be greatly simplified.

Summary of Odd Composites

To summarize the last chapter (chapter 11), all the equations for the elements, sequences and indices into the odd composites, are given by the following formulas (please refer to table 2).

Column 1 Odd Composites

(O5.1) $m(a,j,n) = nP_{j+2} + (P_{j+2}[1-\text{Mod}((P_{j+2}-1)/2, 3)] -1)/6$

(O5.2) $\Theta(a,j,n) = 6m(a,j,n) + 1 = 6nP_{j+2} + P_{j+2}[1-\text{Mod}((P_{j+2}-1)/2, 3)]$

(O5.3) $\{\Theta(a,j,n)\} = \{6m(a,j,n) + 1 \mid j, n = 1, 2, .. \infty\}$

Column 3 Odd Composites

Note that the index into these odd composites is simply the dummy index m.

(O1.1) $\Theta(c,0,m) = 6m + 3 = 3[2m + 1]$

(O1.2) $\{\Theta(c,0,m)\} = \{6m + 3 \mid m = 1, 2, 3, .. \infty\}$

Column 5 Odd Composites

(O8.1) $m(e,j,n) = nP_{j+2} +$

$(P_{j+2}[1-2\text{Mod}((P_{j+2}+1)/2, 3)] +1)/6$

(O8.2) $\Theta(e,j,n) = 6m(e,j,n) - 1 =$

$6nP_{j+2} + P_{j+2}[1-2\text{Mod}((P_{j+2}+1)/2, 3)]$

(O8.3) $\{\Theta(e,j,n)\} = \{6m(e,j,n) - 1 \mid j, n = 1, 2, .. \infty\}$

Note that the only prime in Column 3 is three (3). Columns 2, 4 and 6 contain all the even naturals and therefore do not contain any odd primes.

Possible Catch 22

Note also that the general formulas (equations (O5.2) and (O8.2)) for generating the column 1 and column 5 odd composites assume that the primes (P_{j+2}) are known. Thus, at first sight, a catch 22 situation seems to arise in that any formula that uses these odd composites to identify the primes must already know what the primes are! It turns out that knowing only a first few primes allow for the generation of many composites and by the SOE, generation of many primes. The following section may help clarify the situation.

Generating a Few Primes

Suppose it has been found by hand, that five (5) is the first (j = 1 of P_{j+2}) epsilon prime. The first row of table 3 gives the formulas for the first index and sequence element for the column 1 (referring to table 2) and column 5 odd composites as

a-index	a-comps	e-index	e-comps
5n − 1	30n − 5	5n + 1	30n + 5

which means that the first sequence element for the column 1 odd composites is given by

(PG1) $\Theta(a,1,n) = 30n - 5$

and the first sequence element for the column 5 odd composites is

(PG2) $\Theta(e,1,n) = 30n + 5$

The first column 1 odd composite for n = 1, is

(PG1.1) $\Theta(a,1,1) = 30(1) - 5 = 25$

The first column 5 odd composite for n = 1, is

(PG2.1) $\Theta(e,1,1) = 30(1) + 5 = 35$

Listing all the column 1 members (table 2) starting with seven (7) up to twenty-five (25) exclusive yields the following sequence of non-composites (alpha primes).

(PG3) {7, 13, 19}

Listing all the column 5 members up to thirty-five (35) exclusive yields the following sequence of non-composites (epsilon primes).

(PG4) {5, 11, 17, 23, 29}

This means that the first three alpha primes are

(PG3.1) $\{P(\alpha, j) \mid j = 1, 2, 3\} = \{7, 13, 19\}$

and the first 5 epsilon primes are

(PG4.1) $\{P(\varepsilon, i) \mid i = 1, 2, ..5\} = \{5, 11, 17, 23, 29\}$

Thus, the first eight primes greater than three have been generated. Conventionally, these primes are

(PG5) $\{P_{J+2} \mid J = 1, 2, 3 .. 8\} =$
$\{P_3, P_4, P_5, P_6, P_7, P_8, P_9, P_{10}\} =$
$\{5, 7, 11, 13, 17, 19, 23, 29\}$

The second column 1 odd composite for n = 2, is

(PG1.2) $\Theta(a,1,2) = 30(2) - 5 = 55$

The second column 5 odd composite for n = 2, is

(PG2.2) $\Theta(e,1,2) = 30(2) + 5 = 65$

Since eight primes (including the first epsilon prime) were generated from the first epsilon prime, the first alpha prime was found to be seven (7) or $P(\alpha, 1) = 7$. This prime (j = 2 of P_{j+2}) generates composites given by the second row of table 3 which are

a-index	a-comps	e-index	e-comps
7n + 1	42n + 7	7n − 1	42n − 7

This second row of table 3 gives the formulas for the second index and sequence element for the column 1 (referring to table 2) and column 5 odd composites which means that the second sequence element for the column 1 odd composites is given by

(PG6) $\Theta(a,2,n) = 42n + 7$

and the second sequence element for the column 5 odd composites is

(PG7) $\Theta(e,2,n) = 42n - 7$

The second column 1 odd composite for $n = 1$, is

(PG6.1) $\Theta(a,2,1) = 42(1) + 7 = 49$

The second column 5 odd composite for $n = 1$, is

(PG7.1) $\Theta(e,2,1) = 42(1) - 7 = 35$

which duplicates the first column 5 odd composite. Thus, the next two column 1 odd composites after 25 are 49 and 55. This means that all the column 1 members of table 2 exclusively between twenty-five (25) and forty-nine (49) are alpha primes. These are

(PG8) $\{25+6, 25+2(6), 25 + 3(6)\} = \{31, 37, 43\}$

Similarly, listing all the column 5 members from thirty-five (35) up to sixty-five (65) exclusive yields the following sequence of epsilon primes.

(PG9) $\{41, 47, 53, 59\}$

This means that the next three alpha primes are

(PG8.1) $\{P(\alpha, j) \mid j = 4, 5, 6\} = \{31, 37, 43\}$

and the next four (4) epsilon primes are

(PG9.1) $\{P(\varepsilon, i) \mid i = 6, 7, ..9\} = \{41, 47, 53, 59\}$

Thus, the next seven primes have been generated (by knowing $P_{j+2} = 7$ for $j = 2$). Conventionally, these primes are denoted by

(PG10) $\{P_{J+2} \mid J = 9, 10, 11 .. 15\} =$
$\{P_{11}, P_{12}, P_{13}, P_{14}, P_{15}, P_{16}, P_{17}\} =$
$\{31, 37, 41, 43, 47, 53, 59\}$

Note that by knowing the first epsilon prime (5) has allowed the generation of the following conventional fifteen element prime sequence including the first alpha prime (7).

(PG11) $\{P_{J+2} \mid J = 1, 2, 3 .. 15\} =$
$\{P_3,P_4,P_5,P_6,P_7,P_8,P_9,P_{10},P_{11},P_{12},P_{13},P_{14},P_{15},P_{16},P_{17}\} =$
$\{5,7,11,13,17,19,23,29,31,37,41,43,47,53,59\}$

Therefore, the sieve of Eratosthenes (SOE) has been greatly simplified. Only column 1 and column 5 (referring to table 2) members need be considered.

Equation

(O5.1) $m(a,j,n) = nP_{j+2} + (P_{j+2}[1-\text{Mod}((P_{j+2}-1)/2, 3)] - 1)/6$

for the index into the column 1 odd composites allow these composites to be calculated and eliminated (crossed out). This means that the sequence of new alpha primes is

(PG12) $\{P(\alpha,L) \mid L=L_1 .. L_2\} =$
$\text{Ezsort}\{\{6k+1 \mid k = k_1 .. k_2\}$
$-\text{Ezsort}\{6(m(a,j,n)+1 \mid j = j_{\alpha 1} .. j_{\alpha 2}, n = n_{\alpha 1} .. n_{\alpha 2})\}\}$

Equation

(O8.1) $m(e,j,n) = nP_{j+2} + (P_{j+2}[1-2\text{Mod}((P_{j+2}+1)/2, 3)] + 1)/6$

for the index into the column 5 odd composites allow these composites to be calculated and eliminated (crossed out). This means that the sequence of new epsilon primes is

120 Primal Proofs

(PG13) $\{P(\varepsilon,M) \mid M = M_1 .. M_2\} =$

$$\text{Ezsort}\{\{6q-1 \mid q = q_1 .. q_2\}$$
$$- \text{Ezsort}\{6(m(e,j,n) - 1 \mid j = j_{\varepsilon 1} .. j_{\varepsilon 2}, n = n_{\varepsilon 1} .. n_{\varepsilon 2}\}\}$$

New alpha and epsilon primes are identified and sorted into the standard list of primes via

(PG14) $\{P_{j+2} \mid j = j_1 .. j_2\} =$

$$\text{Ezsort}\{ P(\alpha,L), P(\varepsilon,M) \mid L = L_1 .. L_2, M = M_1 .. M_2\}$$

Recall the Ezsort operator is defined in chapter 3. In equations (PG12) and (PG13), k_1, k_2 q_1, q_2 define the prime search range and $j_{\alpha 1}$, $j_{\alpha 2}$, $n_{\alpha 1}$, $n_{\alpha 2}$, $j_{\varepsilon 1}$, $j_{\varepsilon 2}$, $n_{\varepsilon 1}$, $n_{\varepsilon 2}$ are chosen to generate all composites within the prime search range. In equation (PG14), j_1 and j_2 depend of how many new alpha ($L_1..L_2$) and epsilon ($M_1..M_2$) primes are identified. Thus, more column 1 and column 5 odd composite indices can be calculated resulting in more primes being identified ... ad infinitum.

GLOSSARY

Andrica's Conjecture: The square root of the (n+1)th prime less the square root of the nth prime is always less than unity.

(G1) $\sqrt{P_{n+1}} - \sqrt{P_n} < 1$ where $n \geq 1$ is a natural number.

Bertrand's Postulate: There always exists a prime, p such that

(G2) $n < p < 2n$, where $n > 1$ is a natural number.

Brocard's Conjecture: Between the squares of any two consecutive odd primes, there are at least 4 other primes. In other words

(G3) $\pi(P_{n+1}^2) - \pi(P_n^2) \geq 4$

where $n > 1$ and π is the prime counting function.

Chen Jing-run's Theorem: Every sufficiently large even number is the sum of a prime and another number which is either a prime or has two prime factors.

Coprimality Theorem: Given two primes, P_A and P_B, there exists two naturals N_1 and N_2 such that

(G4) $N_1 P_A - N_2 P_B = 1$

Derichlet's eta function:

(G5) $\eta(s) = \sum_{n=1}^{\infty} (-1)^{n-1}/n^s$

where s is complex. This is sometimes called the alternating zeta function.

Derichlet's Theorem: If A and B are two naturals that share no prime factors, then AX + B generates an infinite number of primes where the natural number X takes on a corresponding infinite number of values.

Euler's Constant:

(G6) $e = \lim_{n \to \infty} (1 + 1/n)^n$

(G6.1) $e = \sum_{n=1}^{\infty} 1/(n-1)!$

Euler's Prime Generating 2nd Degree Polynomial:

(G7) $F(x) = (x-1)^2 + x + 40$

where x is a natural number. For x ranging consecutively from 1 to 40, F(x) produces forty (40) primes!

Euler's Product Formula:

(G8) $\sum_N 1/N^S = \prod_N 1/(1 - 1/P_N^S) = \zeta(s)$

where the sum is over all the naturals, N and the product is over all the primes, P_N. The sum converges for all s > 1 where s is a natural number. Riemann extended Euler's product formula by letting s be a complex number. The resulting function is called the Riemann zeta function. Thus, Euler's product formula is equivalent to the Riemann zeta function when s is a complex number.

Fermat's Little Theorem Generalized: The sum of a set of naturals, $\{n_j \mid j = 1, 2, 3 .. k\}$ raised to some prime power P less the sum of each natural raised to the power P is a natural factor, m of P.

$$(G9) \left(\sum_{j=1}^{j=k} n_j\right)^P - \sum_{j=1}^{j=k} n_j^P = mP$$

Fermat Prime: A prime number which has the form

$$(G10)\ P_F = 1 + 2^{2^N}$$

where P_F is a Fermat prime and N is a natural number.

Fermat's Test for Primeness: If for some natural arbitrary number $N_A < P$ (the alleged prime), the quantity

(G11) $N_A^{(P-1)} - 1$

is evenly divisible by P, then P is a prime.

Fundamental Theorem of Algebra: Every polynomial equation of degree N greater than zero (0) with real or complex coefficients has exactly N real or complex roots.

Fundamental Theorem of Arithmetic: Any Natural number greater than one (1) may be expressed in only one way as

(G12) $N = (P_1)^{M1}(P_2)^{M2}(P_3)^{M3}(P_4)^{M4} .. (P_J)^{MJ}$

where N>1 is any natural number and P_J is the Jth prime and MJ is the exponent of the Jth prime.

Fundamental theorem of calculus:

(G13) $d [\int df] = d[(f(x) + C] = df$

Gamma Function: The gamma function, $\Gamma(s)$ is

(G14) $\Gamma(s) = \int_0^\infty t^{s-1} e^{-t} dt$ where s is complex

Goldbach's Binary Conjecture: Any even natural number greater than two (2) can be expressed as the sum of two (2) primes. This conjecture has been tested for even numbers up to 10^{14}. For details visit www.onderzoekinformatie.nl/en/oi/nod/onderzoek/ OND1256275/

Goldbach's Comet: If Goldbach numbers (see definition below) are plotted against the even naturals, bands of values appear with each band shaped like a comet. A sharp lower band yields the minimal values of Goldbach numbers.

Goldbach Number: The number of ways in which an even number can be expressed as the sum of two primes. See Calvin C. Clawson's *Math Mysteries* Reference.

Goldbach's Ternary Conjecture: Any natural number greater than five (5) can be expressed as the sum of three (3) primes.

Gosper's Factorial Approximation: N factorial is given by

(G15) $N! \approx [(2N + 1/3)\pi]^{1/2} N^N e^{-N}$

and is an improvement over Stirling's factorial approximation in that it yields

(G16) $0! \approx 1.024$

(G17) $1! \approx .996$

Karst's Theorem: The second degree polynomial

(G18) $F_K(X) = 2X^2 - 199$

generates the greatest number of primes for the 1st thousand values of X. X is a natural number.

L'Hopital's rule: If two functions $f(x)$ and $g(x)$ at the point $x = a$ are both zero, i.e. $f(a) = g(a) = 0$, or if $f(a) = g(a) = \infty$ then

(G19) $\lim_{x \to a} f(x)/g(x) = \lim_{x \to a} (df/dx)/(dg/dx)$

Legendre's Conjecture: There exists a prime, P between the square of some natural number, n^2 and the square of the next natural number $(n+1)^2$

(G20) $n^2 < P < (n+1)^2$

Logarithmic Integral, Li(N): The number of primes, $\pi(N)$ less than or equal to some natural number, N (conjectured by Gauss) is given by

(G21) $\pi(N) \approx \int_{2}^{N} dt/Ln(t) = Li(N)$

Glossary 129

Mersenne Prime: A Mersenne prime, P_M is

(G22) $P_M = 2^P - 1$

where P is another prime number.

Polynomial Equation: A polynomial equation of degree N is given by

(G23) $z_N x^N + z_{N-1} x^{N-1} + \ldots + z_1 x + z_0 = 0$

where z is a complex or real constant coefficient

Prime Number Theorem (PNT): The number of primes, $\pi(N)$ less than or equal to some natural number, N (conjectured by Gauss) is given by

(G24) $\pi(N) \approx N/Ln(N)$

Ramanujan's Prime Relationships: In the following equations, P_N stands for the Nth prime.

(G25) $\prod_{N=1}^{\infty} (P_N^2 + 1)/(P_N^2 - 1) = 5/2$

(G26) $\prod_{N=1}^{\infty} (1 + 1/P_N^4) = 105/\pi^4$

(G27) $\sum_{k=1}^{\infty} P_k/e^{kx} \approx Ln(x)/x^2$

Riemann's Extended Zeta Function: The extended zeta function of s (s is complex $\neq 1$) is

(G28) $\zeta(s) = 2^s \pi^{s-1} Sin(\pi s/2) \Gamma(1-s) \zeta(1-s)$

where Sin is the normal trigonometric sine function and Γ is the gamma function defined as

(G29) $\Gamma(s) = \int_0^{\infty} t^{s-1} e^{-t} dt$

Riemann's Hypothesis: All non-trivial zeros of Riemann's zeta function extended (s>0) via Derichlet's eta function, $\eta(s)$ given by

(G30) $\zeta(s) = \eta(s)/(1 - 2^{-(s+1)})$

$\zeta(s)$ lie on a line in the complex plane described by $s^{\pm} = 1/2 \pm iy$ for certain values of $y \neq 0$. Trivial zeros occur at all negative even values of s (–2, –4, –6, –8, ...) and y = 0, as can be seen by inspection of Riemann's extended zeta function above.

Riemann's Prime Counting Function: The number of primes, $\pi(N)$ less than or equal to some natural number, N is given by

$$(G31)\ \pi(N) = R(N) + \sum_{\rho} R(N^{\rho})$$

where ρ is a non-trivial zero of the Riemann zeta function and where R is Riemann's function defined as

$$(G32)\ R(N) = \sum_{n=1}^{\infty} (\mu(n)/n) Li(N^{1/n})$$

and $\mu(n)$ is the Möbius function defined as

(G33) $\mu(n) = 0$, if n has repeated prime factors

(G34) $\mu(n) = 1$, if $n = 1$

(G35) $\mu(n) = (-1)^k$ if n is a product of k distinct primes and Li is the logarithmic integral defined by

(G36) $Li(N) = \int_{2}^{N} dt/Ln(t)$

where $Ln(t)$ is the natural logarithm of t.

Riemann's Zeta Function:

(G37) $\sum_{N} 1/N^S = \prod_{N} 1/(1 - 1/P_N^S) = \zeta(s)$

where the sum is over all the naturals, N and the product is over all the primes, P_N. Here, $s \neq 1$ is a complex number.

Rosser - Schoenfeld Theorem:

(G38) $N/Ln(N) < \pi(N) < 1.25506 N/Ln(N)$

where $N > 16$ (Rosser & Schoenfeld 1962)

Selected Prime Relationships:

(G39) $P_{N+1} < 2P_N$

(G40) $P_N + P_{N+1} > P_{N+2}$

(G41) $P_N * P_M > P_{N+M}$

(G42) $\sum_P 1/P \to \infty$

(G43) $\sum_{P_T} 1/P_T = 1.90216$

where P_T are twin prime members (3, 5, 7, 11, 13, 17, 19, .. etc)

Selected Series:

(G44) $\sum_{n=1}^{\infty} 1/n^2 = \pi^2/6 = \zeta(2)$

(G45) $\sum_{n=1}^{\infty} 1/(2n-1)^2 = \pi^2/8$

(G46) $\sum_{n=1}^{\infty} (-1)^{n+1}/n^2 = \pi^2/12$

(G47) $\sum_{n=1}^{\infty} 1/3^n = 1/2$

(G48) $\sum_{n=1}^{\infty} 1/2^n = 1$

(G49) $\sum_{n=1}^{\infty} 1/n^4 = \pi^4/90 = \zeta(4)$

(G50) $\sum_{n=1}^{\infty} 1/(2n-1) \to \infty$

(G51) $\sum_{n=1}^{\infty} 1/n \to \infty$

(G52) $\sum_{n=1}^{J} 1/n = Ln(J) + .5772157..$

where the Euler-Mascheroni constant is .5772157..

(G53) $\sum_{j=1}^{n} (2j-1) = n^2$

(G54) $\sum_{j=1-n+\sigma}^{\sigma} (2j-1) = n^3$

where

(G55) $\sigma = \sum_{1}^{n} n$

(G56) $\pi^2 = 15 \prod_{N=1}^{\infty} 1/(1 + 1/P_N^2)$

where P_N is the nth prime (by Author in 1999)

Sieve of Eratosthenes: An algorithm which identifies all primes from a sequential test list, L_T of naturals composed of the first prime, $P_1 = 2$ to the greatest natural number, N_G to be tested. The first prime, P_1 is left uncrossed in L_T but all multiples of P_1 are crossed out. The next number that is not crossed out is a prime which is left uncrossed in L_T but all multiples of that prime are crossed out. Proceed in this manner until all numbers in L_T have been crossed out. What numbers remain uncrossed are all the primes in L_T.

Singlet Primes: Two primes which differ by one (1). The only set of prime singlets are {2, 3}.

Sophie Germain Prime: A Sophie Germain prime is of the form

(G57) $P_A = 2P_S + 1$

where P_S is a Sophie Germain prime and P_A is another prime.

Stirling's Factorial Approximation: N factorial is given by

(G58) $N! \approx (2\pi)^{1/2} N^{N+1/2} e^{-N}$

and fails for $N = 0$ since it yields

(G59) $0! \approx 0$

instead of 1. For $N = 1$, it yields

(G60) $1! \approx .9221$

Gosper's approximation (included in this glossary above) is a much better approximation for both $N = 0$ and $N = 1$.

Triplet Primes: Three consecutive primes which differ by two (2). The only set of prime triplets is {3, 5, 7}.

Twin Primes: Two primes which differ by two (2).

Twin Prime Conjecture: There is an infinite number of twin primes (two primes that differ by two (2))

Wilson's Prime Testing Function: Wilson's prime testing function is given by

(G61) $F_W(J) = \lessdot \cos^2(\pi[(J-1)! + 1]/J) \gtrdot$

(G62) If $F_W(J) = 1$

then J is a prime.

(G63) If $F_W(J) = 0$

then J is composite. J >1 is any natural to be tested.

Wilson Prime: A Wilson prime, P_W is a prime such that

(G64) $[(P_W - 1)! + 1]/P_W^2 = N$

where N is a natural number.

Vinogradov's Theorem: Every sufficiently large odd number can be expressed as the sum of three (3) primes.

(G65) $2N - 1 = P_X + P_Y + P_Z$

where $N > 10^{7194}$ (Chen & Wang 1996)

Wilson's Theorem:

(G66) If $[(p-1)! + 1]/p = n$

where the factor, n is some natural number, then p is prime.

Axiomatic Foundations

Natural Numbers

The following five axioms known as the Peano axioms will now be listed and described. These axioms are the fundamental definitions of natural numbers. In turn, the fundamental rules of all mathematics can be deduced from this foundation.

Axiom 1: Unity (1) is a natural number

Unity (1) can be defined as something divided by itself. This implies wholeness. Something must have substance. Nothing (0) has no substance.

Axiom 2: The successor of a natural number is another natural number.

Successor is the next (one more than the previous) number in the sequence of natural numbers.

Axiom 3: No two natural numbers have the same successor.

No same successor implies uniqueness. Each natural number is unique and must have one unique successor in the sequence of natural numbers.

Axiom 4: Unity (1) is not the successor of any natural number.

Successor implies that unity (1) is the beginning of the sequence of natural numbers. This also implies that nothing (0) is not a natural number in the sequence of natural numbers.

Axiom 5: If Unity (1) measures something, and the successor of every number measures something, then it is induced that every number in the sequence of natural numbers measures something.

Successor implies that every natural number inherently contains the same kind of substance. If it can be proven that unity (1) has a certain property and the successor of every number has this property, then it is inductively proven that all numbers in the sequence of natural numbers have this property.

Euclidean Geometry

The following five axioms are due to Euclid and form the foundation of Euclidean geometry.

Axiom 1: Any two points can be joined by a straight line segment.

This implies that any line segment must consist of a minimum of two points.

Axiom 2: Any line segment can be extended indefinitely to form a straight line.

This defines a straight line.

Axiom 3: Given any line segment, a circle can be constructed having the line segment as its radius and one endpoint as its center.

This implies that any radius of a circle consists of a line segment.

Axiom 4: All right angles are equal.

Axiom 5: Through a point not on a given straight line, one and only one line can be drawn that never intersects the given line.

This defines two parallel lines.

Equality

The following four axioms are also due to Euclid and define the meaning of equality.

Axiom 1: Things that equal to the same thing are equal to one another.

If B is equal to A and C is equal to A, then B is equal to C.

Axiom 2: If equals are added to equals, the sums are equal to one another.

If A equals B, then $A + C$ is equal to $B + C$.

Axiom 3: If equals are subtracted from equals, the remainders are equal to one another.

If A equals B, then $A - C$ is equal to $B - C$.

Axiom 4: If one object is coincident with another, then they are equal to one another.

This defines coincidence.

Fundamental Rules of Mathematics

In the following rules, the quantities A, B, C, R, Angle, Arclength, X, Y, x, y all stand for real numbers (see types of numbers after the rules section).

Rules of Algebra

1. **Subtraction operator and equivalence operator** used for the definition of zero (0):

(RA1) $A - A = 0$

where $(-)$ is the subtraction operator and $(=)$ is the equivalence symbol.

2. **Division operator** used for the definition of the unity (1) number:

(RA2) $A/A = 1$, $A \neq 0$

where (/) is the division operator. The (≠) symbol means not equal.

3. **Addition operator** used for the definition of both the **multiplication operator** and the number two (2).

(RA3) A + A = 2*A

(RA3.1) A = 1*A = A*1

where (+) is the addition operator and (*) is the multiplication operator. Note that extending (RA3) by adding more A's (i.e. A + A + A = 3*A) along with (RA3.1) may be used to define all the natural numbers {1, 2, 3, 4, ...}.

4. **Commutative property of addition and multiplication**

(RA4) A + B = B + A

(RA4.1) A*B = B*A

5. Associative property of addition and multiplication

(RA5) $A + (B + C) = (A + B) + C = A + B + C$

(RA5.1) $A*(B*C) = (A*B)*C = A*B*C$

The parenthesis operator (…) means perform whatever operation is inside the parenthesis.

6. Distributive property of multiplication and division over addition

(RA6) $A*(B + C) = A*B + A*C$

(RA6.1) $(A + B)/C = A/C + B/C, \quad C \neq 0$

7. Multiplication used in the definition of exponentiation

(RA7) $A*A = A^2$

where 2 is the exponent of A. An extension of (RA7) may be used for defining greater values of the exponent (i.e. $A*A*A = A^3 = A*A^2$)

8. Exponentiation property

(RA8) $A^B = 1/A^{-B}$, $A \neq 0$

9. Distribution of exponentiation over addition and multiplication

(RA9) $A^B * A^C = A^{B+C}$

(RA9.1) $(A^B)^C = A^{(B*C)} = A^{B*C}$

10. The definition of a Logarithm

(RA10) $Log_A A^B = B$

The logarithm to the base of any number is its exponent. The most common bases are the base 10 (common logs) and the base e (natural logs) where e is Euler's constant.

The distribution property of logarithms to the base C over addition is

(RA10.1) $\text{Log}_C A + \text{Log}_C B = \text{Log}_C (A*B)$

Rules of Trigonometry

Please refer to figure 1 with respect to the trigonometric definitions which follow.

1. The definition of an Angle

(RT1) Angle = Arclength/R

Fundamental Rules of Mathematics 149

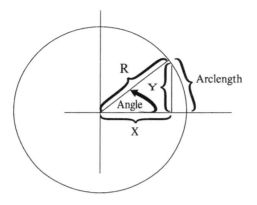

Figure 1

2. Definition of the Sine of an Angle

(RT2) Sin (Angle) = Y/R

3. Definition of the Cosine of an Angle

(RT3) Cos (Angle) = X/R

4. Definition of the Tangent of an Angle

(RT4) Tan (Angle) = Y/X

5. Pythagorean Theorem

(RT5) $X^2 + Y^2 = R^2$

Rules of Hyperbolic Functions

1. Definition of hyperbolic sine (Sinh)

(RH1) $\text{Sinh}(Y) = \frac{1}{2}(e^Y - e^{-Y})$

where e is the base of the natural logarithm.

2. Definition of hyperbolic cosine (Cosh)

(RH1) $\text{Cosh}(Y) = \frac{1}{2}(e^Y + e^{-Y})$

3. Definition of hyperbolic tangent (Tanh)

(RH1) $\text{Tanh}(Y) = \text{Sinh}(Y)/\text{Cosh}(Y)$

Rules of Complex Algebra

1. Definition of i

(RI1) $i = (-1)^{1/2}$

or equivalently

(RI1.1) $i^2 = -1$

2. Definition of a complex number

(RI2) $Z = X + iY$

where Z is a complex number with real part X and imaginary part iY. Both X and Y are real.

3. Definition of the complex plane

If the vertical axis of figure 1 is multiplied by i, then the plane defined by the horizontal axis and the vertical axis is called the complex plane. Any point in the plane has an real X horizontal component and an imaginary Y vertical component which is the geometrical interpretation of (RI2).

4. Euler's complex formula

(RI4) $e^{i*Angle} = \cos(Angle) + i*\sin(Angle)$

where e is Euler's constant. Cos and Sin are the normal sine and cosine of the angle as shown in figure 1. Note that if (RI4) is multiplied by the radius R, then

$$(RI4.1)\ Re^{i*Angle} = R\cos(Angle) + i*R\sin(Angle)$$
$$= R(X/R) + iR(Y/R)$$
$$= X + iY = Z$$

which is exactly the definiton of a complex number given by (RI2).

Rules of Calculus

Definition of the derivative

If $f = f(x)$ is a function of x then its derivative with respect to x is

$$(RC1) \; df/dx = \lim_{\Delta x \to 0} [f(x + \Delta x) - f(x)]/\Delta x$$

where Δx is some interval (change) in x. Note that this can also be thought of as the prescription for calculating the instantaneous slope of f(x) with respect to x.

Definition of the integral

$$(RC2) \; \int df = \int (df/dx)dx = f(x) + C$$

where C is the integration constant. Note that this is a prescription for summing up via the integral ($\int$) the infinitesimal function of f, (df) to reassemble the original function, f(x) plus a constant. Note that the derivative of a constant, C is zero by the definition of a derivative (RC1) since if f(x) = C, then by (RC1)

(RC1.1) $dC/dx = \lim_{\Delta x \to 0} [(C + \Delta C) - (C)]/\Delta x$

$= [C - C]/\Delta x = 0$

since by definition of a constant $(C + \Delta C) = C$ says that constants don't change. Thus, $dC/dx = 0$. Thus, **the fundamental theorem of calculus** is

(RC1.2) $d [\int df] = d[(f(x) + C] = df$

Types of Numbers

There are five types of basic numbers. These are denoted by hollow letters which are:

(TN1) $\mathbb{N} = \{1, 2, 3, \ldots \infty\}$

These are all the natural numbers as defined by the Peano axioms.

(TN2) $\mathbb{Z} = \{-\mathbb{N}, 0, \mathbb{N}\}$

These are all the integers including negative naturals, zero (0) and all the positive naturals.

(TN3) $\mathbb{Q} = \{\{\mathbb{Z}\}, \{N_J/N_K \mid (N_J, N_K) \in \mathbb{Z}, N_K \neq 0\}\}$

These are all the integers and the rationals (integer fractions).

(TN4) $\mathbb{R} = \{\{\mathbb{Q}\}, \{R_I \in \mathbb{Q} \mid I = 1, 2, 3\ldots \infty\}\}$

These are the real numbers including all integers, rationals and irrationals (R_I) that are not (ϵ) integers or rationals.

(TN5) $\mathbb{C} = \{R_J + iR_K \mid (R_J, R_K) \in \mathbb{R} \}$

these are all the complex numbers where $i = -1^{1/2}$.

References

Aczel, Amir D., *The Mystery of the Aleph*, New York, Barnes & Noble, 2005

Ames, Joseph Sweetman & Murnaghan, Francis D., *Theoretical Mechanics An Introduction to Mathematical Physics*, New York, Dover Publications, Inc., 1957

Ball, Keith, *Strange Curves, Counting Rabbits, and other Mathematical Explorations*, Princeton University Press, 2003

Beyer, William H., *CRC Standard Mathematical Tables*, West Palm Beach Florida, CRC Press, Inc., 1972

Bickley, W. G. & Gibson, R. E., *Via Vector To Tensor*, New York, John Wiley & Sons, Inc. 1962

Burnside, William, *Theory of Probability*, New York, Dover Publications, Inc., 1959

Clawson, Calvin C., *Mathematical Mysteries The Beauty and Magic of Numbers*, Cambridge, Massachussetts, Perseus Books, 1999

Derbyshire, John, *Prime Obsession Bernhard Riemann and the Greatest Unsolved Problem in Mathematics*, New York, Penguin Group, 2004

Euclid, *The Elements Book I – XIII*, New York, Barnes & Noble, 2006

Finkbeiner, Daniel T., II, *Introduction to Matrices and Linear Transformations*, San Francisco, W. H. Freeman and Company, 1960

Fulks, Watson, *Advanced Calculus An Introduction to Analysis*, New York, John Wiley & Sons, 1964

Gradshteyn, L. S. & Ryzhik, I. M., *Table of Integrals, Series, and Products*, New York, Academic Press, 1965

Gullberg, Jan, *Mathematics From the Birth of Numbers*, New York, W. W. Norton & Company, 1997

Hawking, Stephen, *God Created the Integers*, London, Running Press, 2005

Hooper, Alfred, & Griswold, Alice L., *A Modern Course in Trigonometry*, New York, Henry Holt & Company, 1959

http://www.exampleproblems.com/wiki/index.php/Riemann_zeta_function

http://mathworld.wolfram.com/RiemannHypothesis.html

http://www.onderzoekinformatie.nl/en/oi/nod/onderzoek/OND1256275/

http://en.wikipedia.org/wiki/Riemann_zeta_function

Kaplan, Robert & Ellen, *The Art of the Infinite*, Oxford, Oxford University Press, 2003

Kells, Lyman M., *Elementary Differential Equations*, New York, McGraw-Hill Book Company, 1960

Korn, Granino A., & Korn, Theresa M., *Mathematical Handbook for Scientists and Engineers*, New York, McGraw-Hill Book Company, 1968

Leithold, Louis, *The Calculus with Analytic Geometry*, New York, Harper & Row, Publishers, 1976

Livio, Mario, *The Equation That Couldn't Be Solved*, New York, Simon & Schuster Paperbacks, 2006

Murdoch, D. C., *Linear Algebra for Undergraduates*, New York, John Wiley & Sons, Inc., 1957

Nahin, Paul J., *An Imaginary Tale, The Story of $\sqrt{-1}$*, Princeton, Princeton University Press, 1998

Nielsen, Kaj L., *College Mathematics*, New York, Barnes & Noble, Inc., 1958

Ogilvy, C. Stanley & Anderson, John T., *Excursions in Number Theory*, New York, Dover Publications, Inc. 1966

Pappas, Theoni, *Mathematical Snippets*, San Carlos, California, Wide World Publishing, 2008

Pappas, Theoni, *The Magic of Mathematics*, San Carlos, California, Wide World Publishing, 1994

Paulos, John Allen, *Beyond Numeracy*, New York, Vintage Books, 1992

Pickover, Clifford A., *A Passion for Mathematics*, New Jersey, John Wiley & Sons, 2005

Pickover, Clifford A., *The Math Book*, New York City, Sterling Publishing, 2009

Polster, Burkard, *Q.E.D. Beauty In Mathematical Proof*, New York, Walker & Company, 2004

Riddle, Douglas F., *Analytic Geometry*, Belmont, California, Wadsworth Publishing Company, 1982

Rockmore, Dan, *Stalking the Riemann Hypothesis*, New York, Random House, Inc., 2005

Rosser and Schoenfeld, *Illinois Journal of Mathematics, 6 (1962),* 64 – 94

Salem, Lionel, Testard, Frederic, Salem, Coralie, *The Most Beautiful Mathematical Formulas*, New York, John Wiley & Sons, Inc., 1992

Sautoy, Marcus Du, *The Music of the Primes*, New York, Perennial, 2003

Sondheimer, Ernst & Rogerson, Alan, *Numbers and Infinity*, Dover Publications, Inc., Mineola, New York, 2006

Sutton, Daud, *Platonic & Archimedean Solids*, New York, Walker and Company, 2002

Thomas, George B., Jr., *Calculus and Analytic Geometry Functions of One Variable*, Reading, Massachusetts, Addison-Wesley Publishing Company, Inc., 1961

Thomas, George B., Jr., *Calculus and Analytic Geometry Vectors and Functions of Several Variables*, Reading, Massachusetts, Addison-Wesley Publishing Company, Inc., 1961

Watkins, Matthew, *Useful Mathematical and Physical Formulae*, New York, Walker & Company, 2001

Wills, A. P., *Vector Analysis With an Introduction to Tensor Analysis*, New York, Dover Publications, Inc., 1958

Wylie, C. R., Jr., *Advanced Engineering Mathematics*, New York, McGraw-Hill Book Company, Inc., 1960

Young, Hugh D., *Statistical Treatment of Experimental Data*, McGraw-Hill Co., Inc., 1962

INDEX

A

Addition operator, 145
alpha (α) primes, 83
alpha prime, 56, 84, 87, 96, 102, 115, 118
alpha primes, 90, 91, 97, 103, 107, 114, 117
Alpha primes, 83
Andrica's conjecture, 47
Andrica's Conjecture, 47, 121
Associative property, 146

B

Bertrand's Postulate, 121
Big Bang, iii
binary conjecture, 19, 20, 37
Binary Conjecture, 11, 36
binary Goldbach conjecture, 19
Brocard's conjecture, 51, 54
Brocard's Conjecture, 51, 121

C

Chen Jing-run's Theorem, 122
coincident, 143
Commutative property, 145
complex numbers, 156
complex plane, 131
Coprimality Theorem, 122

D

Definition of a complex number, 151
definition of an Angle, 148
Definition of hyperbolic cosine, 150
Definition of hyperbolic sine, 150
Definition of hyperbolic tangent, 150
Definition of i, 151
Definition of the complex plane, 151
Definition of the Cosine, 149
Definition of the derivative, 153
Definition of the integral, 153
Definition of the Sine, 149
Definition of the Tangent, 149
definiton of a complex number, 152
Derichlet's eta functiion, 7
Derichlet's eta function, 122, 130
Derichlet's Theorem, 122
derivative of a constant, 154
Diophantine equation, 94, 100
Distribution of exponentiation, 147
distribution property of logarithms, 148
Distributive property, 146
Division operator, 144

E

entanglement, 90, 107
Entanglement, 90
epsilon (ε) primes, 83
epsilon prime, 56, 84, 87, 96, 102, 112, 115, 118
epsilon primes, 83, 90, 91, 97, 102, 107, 114, 117, 119
Equality, 142
equivalence operator, 144

Euclid, 79
Euclidean Geometry, 141
Euler-Mascheroni constant, 134
Euler's complex formula, 152
Euler's constant, 123, 152
Euler's product formula, 123
Euler's Product Formula, 123
even natural, 2, 18, 126
even natural number, 2, 33, 34, 126
even naturals, 86, 112
Even numbers, 20
exponentiation, 146
Exponentiation property, 147
extended zeta function, 69
Extended Zeta Function, 130

F

Fermat prime, 124
Fermat Prime, 124
Fermat's Little Theorem, 124
Fermat's Test for Primeness, 125
frequency of primes, 12
Fundamental Rules of Mathematics, 144
Fundamental Theorem of Algebra, 125
Fundamental Theorem of Arithmetic, 125
fundamental theorem of calculus, 126, 154

G

gamma function, 69, 130
Gamma Function, 126
Goldbach binary conjecture, 11
Goldbach Number, 127
Goldbach's binary conjecture, 11, 20, 31, 61

Goldbach's Binary Conjecture, 126
Goldbach's Comet, 126
Goldbach's ternary conjecture, 36
Goldbach's Ternary Conjecture, 33, 127
Gosper's Factorial Approximation, 127

H

hardware, v
hyperbolic cosine, 70
Hyperbolic Functions, 150
hyperbolic sine, 70

I

infinitesimal, 153
integers, 155
integral, 153
integration constant, 153

K

Karst's Theorem, 127

L

Laurent series, 7, 74
Legendre's conjecture, 49
Legendre's Conjecture, 49
Legendre's Conjecture, 128
L'Hopital's rule, 6, 59, 80, 128
line segment, 141
Logarithm, 147
logarithmic integral, 11, 132

M

Mersenne prime, 129
Mersenne Prime, 129
Möbius function, 131
multiplication operator, 145

N

natural logarithm, 2
natural number, 2, 123, 127, 128, 129, 131, 138, 139
natural numbers, 1, 3, 42, 139, 140, 141, 155
Natural Numbers, 139
non-trivial zeros, 71
non-trivial zeros of the zeta function, 79

O

odd composite, 21, 105, 113, 115, 116, 117
Odd composite, 93
odd composites, 25, 26, 90, 91, 92, 95, 96, 98, 100, 102, 104, 110, 113, 116, 119
Odd composites, 93, 99
Odd Composites, 91, 93, 99
odd natural number, 2, 34
odd naturals, 12, 16, 18, 24
odd primes, 13, 15, 16, 23, 25, 27, 33, 37, 62
odds, 25

P

parallel lines, 142
Peano axioms, 139
Polynomial, 129
prime counting function, 4, 26

Prime Entanglement, 83
prime number, 2
Prime Number Generator, 108
prime number theorem, 4, 5, 6, 8, 9, 11, 58, 80, 81
Prime Number Theorem, 11, 128, 129
prime number theorem (PNT), 54
prime numbers, 108
Prime relationships, 132
prime singlets, 47, 135
prime triplets, 42, 137
Prime Triplets, 42
Proof of Goldbach's Binary Conjecture, 20
Pythagorean Theorem, 150

Q

quantum mechanics, v

R

Ramanujan's Prime Relationships, 129
real number, 3
real numbers, 2, 156
Riemann hypothesis, 68
Riemann Hypothesis, 68, 69, 76
Riemann zeta function, 73, 124, 131
Riemann's Extended Zeta Function, 6
Riemann's Hypothesis, 130
Riemann's Prime Counting Function, 131
Riemann's zeta function, 130
Riemann's Zeta Function, 132
Rosser - Schoenfeld (RS) theorem, 51, 54
Rosser - Schoenfeld (RS) Theorem, 5
Rosser - Schoenfeld Theorem, 132
Rules of Algebra, 144

Rules of Calculus, 153
Rules of Complex Algebra, 151
Rules of Trigonometry, 148

S

Selected Series, 133
sieve of Eratosthenes, 109
Sieve of Eratosthenes, 108, 135
sine function, 69
singlet primes, 46
Singlet Primes, 135
Singlets, 42
smallest odd composite, 23
Sophie Germain prime, 136
Sophie Germain Prime, 135
Stieltjes constants, 7
Stirling's Factorial Approximation, 136
Subtraction operator, 144

T

ternary conjecture, 33
triplet, 44, 45, 46
Triplet Primes, 137
Twin Prime, 56
Twin Prime Conjecture, 55, 58, 64, 137
twin primes, 55, 56, 59
Twin Primes, 137
twins, 59, 60

U

Unlimited Primes, 68

V

Vinogradov's Theorem, 138

W

Wilson prime, 138
Wilson Prime, 138
Wilson's Prime Testing Function, 137
Wilson's Theorem, 138

Z

zeroth prime, 21
zeta function, 68

Made in the USA
Monee, IL
27 April 2021

67064541R00108